# MACHINE LEARNING FOR EVERYONE

## From Basics to Intelligent Systems

AF808176

Arshad Khan

Published by: Khan Consulting & Publishing LLC
Printed in the United States of America
ISBN: 978-0-9660863-8-6

# Preface

The world today runs on data, and the ability to learn from that data has become the defining skill of our time. Machine learning, once confined to research labs and specialized teams, now powers everyday experiences: from the recommendations we see and the assistants we speak to, to the medical diagnoses and financial models that shape modern life. Yet for many, the inner workings of these systems remain mysterious, wrapped in technical jargon and mathematical abstraction.

*Machine Learning for Everyone* was written to close that gap. This book is designed for readers who may not come from a computer science or data engineering background but want to understand how intelligent systems are built, trained, and deployed. Whether you are a business professional, student, educator, or simply a curious learner, this book will guide you step by step – from the core ideas that drive machine learning to the techniques that make systems adaptive, reliable, and ethical.

Over the years, I have taught and consulted with audiences ranging from executives and analysts to engineers and educators. In each setting, one truth became clear: machine learning is not just a technical discipline – it is a way of thinking about problems, patterns, and prediction. My goal in this book is to make that way of thinking accessible. You will find clear explanations and logically structured discussions rather than dense mathematics or code-heavy examples. The emphasis is on conceptual understanding–helping readers grasp why machine learning works the way it does, not just how it does.

As a program manager for an AI initiative in Silicon Valley and a long-time analytics consultant, I have seen firsthand how organizations transform when data-driven intelligence is applied thoughtfully. That perspective informs the practical tone of this book—bridging theory with real-world implementation.

The book progresses logically: we begin with the foundations of artificial intelligence, move through the principles of learning from data, and explore the algorithms and processes that bring models to life. Later chapters address the realities of deploying, testing, and governing machine learning systems responsibly, because understanding how these systems impact people and society is just as important as knowing how they work.

If you finish this book with a clear, confident grasp of what machine learning is, how it learns, and how it can be used responsibly in practice, then it will have served its purpose. I would also like to express my gratitude to Amna Mahzooz for her thoughtful review of the manuscript and her valuable feedback.

I invite you to explore, question, and think critically about the systems shaping our digital world. The future of intelligent technology should not belong only to programmers and data scientists – it should be understood by everyone.

*Arshad Khan*

# Table of Contents

# Glossary of Key Terms

| | |
|---|---|
| **Activation Function** | A mathematical operation applied to the output of neurons in a neural network to introduce non-linearity, allowing the model to learn complex patterns. |
| **Actor-Critic Methods** | A reinforcement learning approach combining value estimation ("critic") and policy optimization ("actor") to improve stability and learning efficiency. |
| **A/B Testing** | A controlled experiment that compares two versions of a model or feature (A and B) to determine which performs better under real-world conditions. |
| **Autoencoder** | A neural network designed to compress data into a lower-dimensional representation and then reconstruct it, commonly used for dimensionality reduction and anomaly detection. |
| **Batch Normalization** | A training technique that normalizes layer inputs across mini-batches, stabilizing learning and mitigating issues like gradient explosion or vanishing. |

| | |
|---|---|
| **Bias-Variance Trade-off** | The balance between a model being too simple (high bias) or too complex (high variance), which determines how well it generalizes to new data. |
| **Cross-Validation** | A model-evaluation method that repeatedly splits data into training and testing sets to obtain more reliable performance estimates. |
| **Data Augmentation** | The process of generating additional training examples by modifying existing data, such as rotating images or rephrasing text, to improve generalization. |
| **Data Leakage** | When information from the validation or test set unintentionally influences the training process, leading to unrealistically high-performance results. |
| **Decision Tree** | A model that predicts outcomes through a sequence of decision rules structured like a flowchart; valued for interpretability but prone to overfitting. |
| **Diffusion Models** | Generative models that learn to create realistic data, such as images or audio, by iteratively denoising random noise. |
| **Ensemble Methods** | Techniques that combine multiple models to produce more accurate and stable predictions, examples include Random Forest and Gradient Boosting. |
| **Feature Engineering** | The process of creating or transforming input variables to make patterns more discoverable to machine-learning algorithms. |

**Gradient Descent** — The foundational optimization algorithm that adjusts model parameters in the direction that minimizes prediction errors.

**Huber Loss** — A loss function for regression that combines the advantages of mean-squared and mean-absolute errors, making it more robust to outliers.

**Hyperparameters** — Configuration settings (e.g., learning rate, tree depth) that control the training process but are not learned directly from data.

**Loss Function** — A mathematical measure of how far model predictions deviate from actual outcomes; the central objective minimized during training.

**MLflow / DVC (Data Version Control)** — Tools for experiment tracking and dataset versioning that ensure reproducibility in machine-learning workflows.

**MLOps** — The integration of machine-learning practices with DevOps principles, enabling continuous delivery, automated testing, and scalable deployment of AI systems.

**Overfitting** — When a model performs well on training data but poorly on new data because it memorized noise instead of learning underlying patterns.

**Principal Component Analysis (PCA)** — A technique that reduces data dimensionality by transforming variables into orthogonal components capturing maximum variance.

**Q-Learning**
A reinforcement learning algorithm that learns optimal actions by estimating the expected cumulative reward for each state-action pair.

**Regularization**
A collection of techniques, such as L1/L2 penalties or dropout, that limit model complexity to prevent overfitting.

**Reinforcement Learning (RL)**
A learning paradigm where an agent interacts with an environment, receiving rewards or penalties to learn optimal actions.

**Residual Connections (ResNets)**
Neural network shortcuts that let information bypass certain layers, improving gradient flow and training efficiency in deep architectures.

**Semi-Supervised Learning**
A hybrid learning approach that uses a small amount of labeled data along with a larger pool of unlabeled data to improve performance.

**Stochastic Gradient Descent (SGD)**
A variant of gradient descent that uses small random subsets (mini-batches) of data to accelerate training.

**Transformer**
A deep-learning architecture that uses attention mechanisms to capture long-range dependencies in sequential data such as text or speech.

**Underfitting**
When a model is too simple to learn the underlying structure of data, resulting in poor accuracy on both training and test sets.

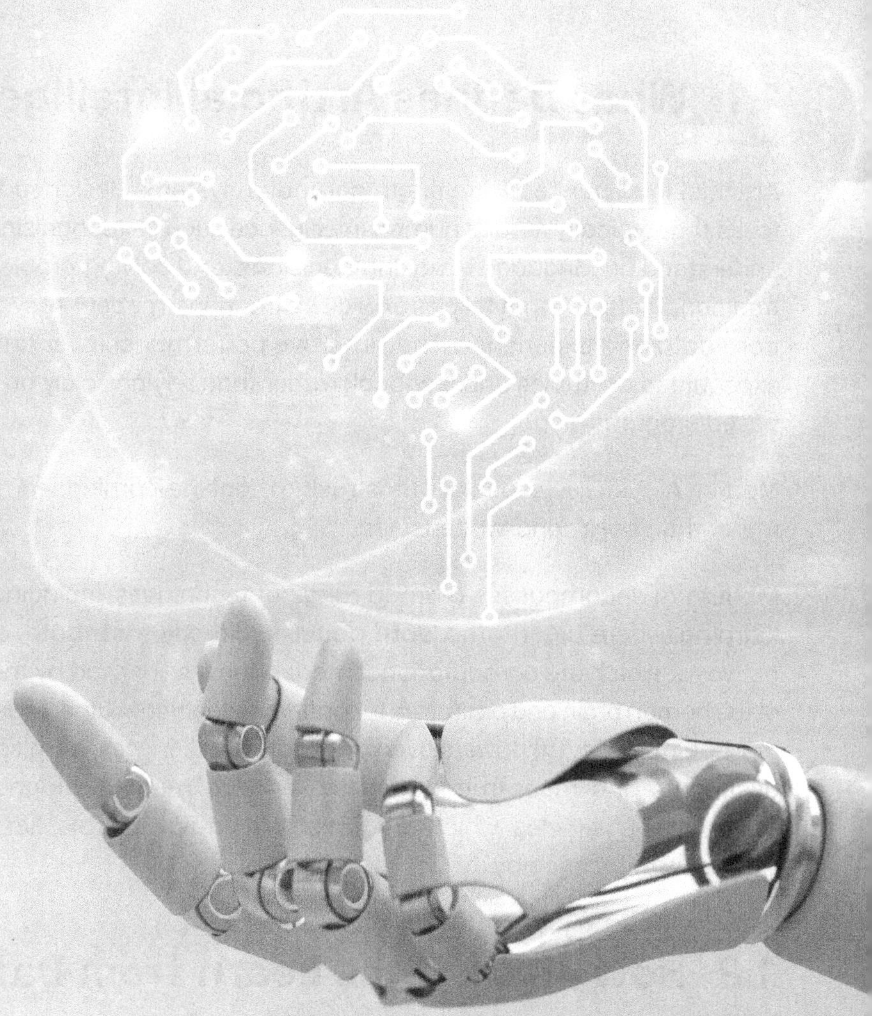

Chapter 1

# Introduction to Artificial Intelligence

# 1.1 What Defines Artificial Intelligence?

Artificial Intelligence (AI) refers to computer systems designed to perform tasks that typically require human intelligence such as recognizing patterns, understanding languages, making decisions, and solving problems. Unlike traditional software that follows explicit instructions, modern AI systems learn from data and experience. They improve performance over time through exposure to examples and feedback rather than relying solely on fixed, rule-based programming.

Modern AI systems, especially those using machine learning, learn from data rather than from hand-written rules.

Modern AI encompasses a broad range of techniques, including machine learning, where algorithms learn patterns directly from data, and neural networks, which are computational models loosely inspired by the structure of the human brain and optimized for pattern recognition rather than biological replication. Today, AI powers everyday tools and services, from smartphone assistants and recommendation systems to medical diagnostics and autonomous vehicles, which are transforming how people live, work, and interact with technology.

# 1.2 How AI Systems Learn from Data

At the heart of AI lies the idea of learning from experience. The process by which an AI system develops its "intelligence" can be summarized in several key steps, described below.

**Learning from Examples**

An AI system begins by analyzing large numbers of examples. For instance, if the goal is to recognize birds, the system is trained using thousands of images labeled "bird" and "not bird". Over time, exposure to diverse examples helps the system understand what truly defines each category rather than memorizing specific pictures.

### Finding Patterns

Through analysis, AI identifies patterns and relationships. It notices, for example, that birds often have feathers, wings, and distinctive body shapes. These repeated features become the clues that the system uses to recognize similar objects in new situations.

### Making Predictions

Once trained, AI can examine new, unseen images and predict whether they contain a bird, applying the patterns it has previously learned. Each prediction reflects the model's best estimate based on everything it has learned from its training data.

### Getting Feedback

When its predictions are evaluated, AI adjusts itself based on accuracy, reinforcing correct responses and modifying errors. If it confuses a dog for a wolf, it refines its internal understanding to improve future performance. This feedback loop is what transforms trial and error into genuine learning.

### Continuous Improvement

This cycle repeats with every new piece of data and feedback, allowing AI to evolve over time and become increasingly capable. The more data it processes, the more refined and reliable its understanding of the world becomes.

This process closely mirrors how humans learn. Just as a child improves their ability to recognize animals by seeing many examples, receiving corrections, and practicing, AI systems refine their understanding by processing more data and learning from mistakes.

## 1.3 Core Capabilities of AI

AI now underpins a wide range of applications across industries and daily life. Its capabilities include:

- **Visual Recognition**

  Identifying and classifying objects, faces, and scenes in images and videos, supporting uses like facial recognition, quality control, and medical image analysis.

- **Language Understanding and Generation**

  Comprehending and producing human language for translation, summarization, chatbots, and virtual assistants.

- **Speech Processing**

  Converting speech to text, recognizing voice commands, and generating natural-sounding responses.

- **Prediction and Forecasting**

  Using historical data to forecast outcomes in finance, healthcare, logistics, and weather.

- **Decision-Making**

  Evaluating complex information to make choices such as credit approval, medical diagnosis, or route optimization.

- **Recommendation and Personalization**

  Suggesting relevant products, entertainment, or information based on user behavior.

- **Automation**

  Enabling robots and digital systems to perform repetitive or physical tasks autonomously.

- **Anomaly Detection**

  Recognizing unusual patterns to flag potential fraud, security breaches, or equipment failures.

In real-world applications, these capabilities often work together. For example, an AI system may combine visual recognition, natural language understanding, and prediction models to solve complex, integrated problems. Generative AI extends traditional analytic capabilities by enabling systems to create new content text, images, or code based on learned patterns, marking a shift from analysis to creation.

## 1.4 Categories of Artificial Intelligence

AI can be broadly categorized into three conceptual types:

- **Artificial Narrow Intelligence (ANI):** AI designed for specific tasks, such as image recognition or language translation. Nearly all current AI systems fall into this category.
- **Artificial General Intelligence (AGI):** A theoretical form of AI capable of performing any intellectual task a human can do. AGI is still a goal for the future, not something that exists today.
- **Artificial Superintelligence (ASI):** A hypothetical stage where AI surpasses human intelligence in every area. Though speculative, it raises key questions about control, safety, and societal impact.

## 1.5 Why Machine Learning Drives Modern AI

Although "artificial intelligence" is a broad term covering many techniques, machine learning has emerged as its driving force. Machine learning provides a practical mechanism through which computers extract patterns from data and improve with experience. It is the foundation, from product recommendation systems and fraud detection to autonomous vehicles and generative tools.

In the next chapter, we will explore how machine learning works: how computers learn from examples, how they make predictions, and how they continuously refine their understanding, all of which form the core of modern intelligent systems.

Chapter 2

# Fundamentals of Machine Learning

## 2.1 How Machine Learning Fits into AI

Artificial Intelligence represents the broader vision of creating systems that can think and act intelligently. However, it is machine learning that turns this vision into reality. Machine learning enables computers to learn directly from data, discovering patterns and relationships without being explicitly programmed for every rule or outcome. Instead of coding solutions by hand, the system is given examples and learns how to solve the problem on its own. This data-driven approach has transformed computing, allowing machines to handle tasks that once required human judgment, such as recognizing speech, understanding images, translating languages, and playing complex games.

In this chapter, we explore the fundamental principles of machine learning, explaining how it differs from traditional programming, how models are trained and generalized, and why it powers nearly every modern AI application. Understanding these foundations provides insight, from recommendation engines to autonomous vehicles, are built, improved, and deployed.

## 2.2 Core Principles of Machine Learning

At its core, machine learning concerns algorithms that improve automatically through experience. The learning process involves exposure to data, detection of patterns or regularities, and adjustment of internal parameters to represent those patterns. The goal is generalization, performing well on new, unseen data rather than merely memorizing training examples.

Machine learning operates under the principle that systems can learn relationships between inputs and outputs directly from examples. With enough data and computing power, these systems can learn complex behaviors that fixed rules cannot capture. This capacity to generalize from experience forms the foundation of nearly all modern AI systems.

## 2.3 The Machine Learning Workflow

Traditional programming relies on hand-crafted rules that map inputs to outputs. Machine learning reverses this approach by letting the system learn the rules from data. Instead of being told the rules, the system learns them from experience—given inputs and desired outputs (or feedback about them), it infers the underlying function or policy that connects the two. This data-driven approach excels when rules are unknown, too complex to specify, or constantly changing with context.

The machine learning process typically unfolds through several stages:

1. **Data Collection:** Gathering relevant examples for the learning task.
2. **Data Preprocessing:** Cleaning, transforming, and organizing data for use by algorithms.
3. **Model Selection:** Choosing an appropriate algorithm or model architecture based on the problem type.
4. **Training:** Adjusting the model's internal parameters to minimize errors in the training data.
5. **Validation:** Testing the model on held-out data to fine-tune hyperparameters and prevent overfitting.
6. **Testing:** Evaluating final model performance on entirely new data.
7. **Deployment and Monitoring:** Implementing the trained model in production and tracking performance over time.

This cycle mirrors human learning: observation, practice, feedback, and improvement. Continuous monitoring ensures that models remain accurate as conditions or data patterns evolve.

# 2.4 Foundational Concepts in Machine Learning

Understanding several key concepts is essential for grasping how machine learning works in practice. These ideas form the foundation for how models learn from data, make predictions, and improve over time. By clarifying terms such as features, labels, training, and evaluation, readers can better appreciate how each component contributes to building intelligent systems.

## Core Building Blocks of Learning

This subsection focuses on the essential elements that define how a model is built, trained, and improved. It explains what data the model learns from, what it aims to predict, and how it refines itself over time.

### Features

These are the measurable attributes or pieces of information that a model uses to learn. In a photo, features might include colors, textures, or shapes; in a loan application, income and credit history. Choosing and preparing useful features–called feature engineering–is vital for good models, though deep learning often learns features automatically. Well-chosen features make patterns more apparent to the algorithm, improving both accuracy and interpretability. Poor or irrelevant features, on the other hand, can confuse the model and lead to weak or biased predictions.

### Labels

These are the known outcomes used to train a model in supervised learning. For an email classifier, labels could be "spam" or "not spam." In sentiment analysis, for example, labels might indicate whether a customer review is positive, negative, or neutral. The accuracy of labels directly affects the reliability of the model's predictions, and mislabeled data can lead to systematic errors that distort learning outcomes. High-quality labeling often requires human judgment or domain expertise, making it one of the most time-consuming yet crucial parts of building a successful model.

### Model

A model is the system that learns patterns and relationships between input data and the correct answers. It can be as simple as a straight-line formula or as complex as a large network that mimics how information flows in the brain. The goal is to build a model that understands real patterns in the data without being misled by random noise. Once trained, the model serves as a mathematical representation of knowledge that can generalize to new, unseen situations, making it a reusable engine for prediction and decision-making.

### Loss Function

The loss function tells the model how well it is doing, like a report card that measures its mistakes. It compares the model's predictions to the correct answers and gives a score showing how far off they are. A lower score means the model is performing better. Different tasks use different loss functions, such as squared error for predicting numbers and cross-entropy for classifying categories. Therefore, selecting the right one is essential for effective learning.

Together, these components describe not only how machine learning models learn but also how their performance and reliability are ensured once deployed. Having explored these key elements of learning, the next step is to understand how models maintain accuracy and trustworthiness in practice.

## Performance and Reliability in Practice

This subsection explores how well a trained model performs on unseen data and what practices ensure its results are both trustworthy and generalizable.

### Optimization

Optimization is the process of improving the model by adjusting its parameters to minimize the loss. Algorithms such as gradient descent take iterative steps in the direction that reduces error, gradually refining the model to achieve better predictions. Common optimizers such as *Adam, RMSProp, and Adagrad* adapt learning rates dynamically to accelerate convergence. This process continues until the model reaches a point where further improvements are minimal, striking a balance between accuracy, efficiency, and generalization to new data.

### Generalization

Generalization is a model's ability to perform well on new, unseen data. A model that merely memorizes the training examples will fail to generalize beyond them. Achieving the right balance between underfitting (too simple) and overfitting (too complex) depends on managing the bias-variance trade-off. Models with high bias tend to underfit by oversimplifying patterns, while models with high variance overfit by capturing noise instead of meaningful relationships. Striking this balance is central to building models that learn effectively and make reliable predictions.

### Evaluation Metrics

Evaluation metrics measure how well a model performs. The right metric depends on the specific goal: for a spam filter, precision (avoiding false positives) may be most important, while in medical screening, recall (avoiding false negatives) often matters more. These metrics turn model performance into clear, meaningful measures of success and guide improvements during development.

### Avoiding Data Leakage

A common pitfall in model development is data leakage, when information from the validation or test sets unintentionally influences the training process. This can make a model appear more accurate than it truly is. Preventing leakage means keeping datasets strictly separate and applying the same preprocessing to each, so evaluation reflects real-world performance on new data. Having established the foundational concepts of how machine learning models learn and improve, we now turn to their real-world impact—how these systems are applied across domains to perceive, predict, create, and discover.

# 2.5 Real-World Applications of Machine Learning

Machine learning has moved from research labs into nearly every sector of society. Its capabilities can be grouped into several core categories that together illustrate the breadth of what learning systems can achieve. From healthcare and finance to transportation and entertainment, machine learning now powers decision-making, automation, and personalization at scale. These applications demonstrate how data-driven intelligence transforms both everyday experiences and complex industrial processes.

## Perception and Prediction

### Pattern Recognition: Teaching Machines to Perceive and Understand

This fundamental capability allows systems to categorize information, identify regularities, and spot anomalies, forming the basis of modern perception systems. Machine learning excels at finding patterns that are too subtle or complex for explicit programming. Its impact can be seen across many familiar technologies, such as the following:

- **In Our Daily Lives:** We encounter pattern recognition in technologies like facial recognition that unlocks our phones, computer vision that enables augmented reality filters, and speech recognition that accurately transcribes voice commands for digital assistants.
- **Transforming Industries:** This capability drives crucial applications across sectors. In healthcare, it analyzes medical images to detect tumors, fractures, and other anomalies. In finance, it powers real-time fraud detection by identifying unusual transaction patterns. Cybersecurity systems use it to identify network intrusions and classify malware, while manufacturing employs it for automated quality control.

### Prediction: Forecasting Future Outcomes

By learning from historical data, machine learning models can make informed forecasts about future events and behaviors. This predictive capability enables proactive decision-making across numerous domains. Its applications are visible in both everyday experiences and large-scale industry operations:

- **In Our Daily Lives:** Recommendation systems on streaming services and e-commerce platforms predict what content or products users will prefer. Search engines use learned ranking models to deliver the most relevant results based on predicted relevance.
- **Transforming Industries:** Prediction drives business optimization through demand forecasting in retail and predictive maintenance in manufacturing, which anticipates equipment failures before they occur. In finance, it enables algorithmic trading and credit risk assessment, while weather services use it for increasingly accurate forecasts.

## Creation and Discovery

### Generation: Creating New Content and Solutions

Going beyond analysis, generative models learn the underlying structure of data so thoroughly they can produce new, original content that resembles the training data. This represents one of machine learning's most revolutionary capabilities, as illustrated in the examples below:

- **In Our Daily Lives:** Generative AI can create realistic images from text descriptions, compose music, write coherent text, and power advanced conversational assistants.
- **Transforming Industries:** In healthcare and pharmaceuticals, generative models help design new drug molecules and predict molecular properties. Research scientists use them for materials discovery, while creative industries employ them for content creation and design assistance.

## Decision-Making: Optimizing Complex Strategies and Actions

This form of intelligence allows systems to not just predict outcomes but to choose actions that lead to the best possible results over time. Its impact can be seen in both everyday conveniences and large-scale industrial applications:

- **In Our Daily Lives:** This technology powers the recommendations for what video to watch next, optimizes travel routes in real-time on navigation apps, and enables AI to achieve superhuman performance in complex strategic games like chess and Go.
- **Transforming Industries:** Decision-making algorithms are crucial for autonomous vehicle navigation and lane-changing decisions. In robotics, they enable machines to learn complex manipulation and locomotion skills. Resource management applications include optimizing data center cooling, energy grid management, and supply chain logistics.

## Discovery and Exploration: Finding Hidden Patterns and Structures

Machine learning can uncover hidden relationships and simplify complex data, serving as a powerful tool for scientific discovery and data exploration. This capability helps humans understand and work with increasingly complex datasets, as the following examples show.

- **In Our Daily Lives:** Recommendation systems on streaming and shopping sites use these techniques to discover your hidden tastes and suggest new movies or products. Personal finance apps use them to categorize your spending and identify unusual patterns for fraud detection.
- **Transforming Industries:** In scientific research, machine learning identifies new categories of galaxies, groups genes with similar expression patterns, and aids in protein folding prediction. Businesses use it for customer segmentation based on purchasing behavior, while across industries it helps simplify complex data for visualization and the discovery of meaningful patterns.

The reach of machine learning continues to expand as methods improve, and more data becomes available. These interconnected capabilities–recognizing patterns, predicting outcomes, generating content, making decisions, and discovering hidden structures–together solidify machine learning's role as a foundational technology that is transforming how we work, live, and understand our world.

While machine learning applications vary widely, they are all built upon a few fundamental learning paradigms. The next section explores these core types–supervised, unsupervised, and reinforcement learning–each representing a distinct way machines acquire knowledge from data.

## 2.6 Categories of Machine Learning Approaches

Machine learning systems differ in how they receive information and feedback during learning. The three main paradigms are supervised, unsupervised, and reinforcement learning, each reflecting a distinct way of acquiring knowledge. Understanding these paradigms provides the foundation for recognizing how different algorithms learn, adapt, and make decisions. Their characteristics are described below:

### Supervised Learning

The system learns from labeled datasets containing both inputs and correct outputs. It tries to predict outcomes for new inputs by learning the mapping between features and labels. Examples include email spam detection, credit-risk scoring, and medical image classification.

## Unsupervised Learning

The system explores unlabeled data to find patterns or groupings without explicit guidance. It discovers structure rather than predicting known outcomes. Examples include customer segmentation, topic modeling, and anomaly detection.

## Reinforcement Learning

The system learns by interacting with an environment, receiving feedback in the form of rewards or penalties. Over time, it develops strategies that maximize long-term reward. Examples include game-playing AI, robotic control, and traffic-signal optimization.

Each learning type supports different kinds of intelligence: supervised learning predicts, unsupervised learning discovers, while reinforcement learning decides and acts.

Chapter 3

# Learning Paradigms

## — Supervised, Unsupervised, and Reinforcement Learning

## 3.1 Overview of Machine Learning Paradigms

Machine learning systems differ in how they acquire knowledge from data. The learning paradigm defines the nature of the feedback or guidance available during training and determines how the system refines its internal model. Machine learning generally operates under three paradigms: supervised, unsupervised, and reinforcement learning. Each represents a different way of learning from experience—similar to how humans learn with guidance, by exploring, or by receiving feedback from their environment.

Hybrid paradigms—semi-supervised and self-supervised learning—bridge labeled and unlabeled data, improving efficiency in real-world scenarios. Self-supervised learning and semi-supervised learning differ mainly in how labels are generated. Self-supervised methods create labels automatically from the data itself, while semi-supervised methods use a small set of human-labeled examples combined with a larger pool of unlabeled data.

## 3.2 Supervised Learning Explained

Supervised learning forms the backbone of most practical machine learning applications today, much like studying from a textbook that includes the answers. The algorithm is trained on a dataset where every input (like a picture or a piece of text) is paired with a known output, called a label or target. The system's goal is to learn the general rule that maps inputs to their correct outputs so it can accurately predict labels for new, unseen data.

## The Foundation: Data and Labels

The process hinges on the quality and quantity of the labeled data:

- Labels define what the model should predict: in classification, they are categories (like 'spam' or 'cat'), and in regression, they are numbers (like income or temperature).
- Creating these datasets often requires significant human effort to annotate data, and the model's performance is fundamentally constrained by the accuracy and completeness of these labels.

## The Learning Process: Training a Model

A model is a mathematical function that the algorithm adjusts during training. The process follows a clear cycle:

**Make a Prediction**

The model takes a training example and produces an output.

**Measure the Error**

A loss function calculates how far the prediction is from the correct answer.

**Learn from the Mistake**

An optimization algorithm (most commonly a form of gradient descent) calculates how to tweak the model›s internal parameters to reduce the error.

**Repeat**

This cycle runs for multiple passes (epochs) over the training data until the model›s performance stops improving.

## The True Test: Evaluation and Generalization

The ultimate goal is generalization—performing well on new, unseen data. To ensure this, the model is evaluated on a separate test set that was held out during training. Robust techniques like cross-validation provide a true measure of its capabilities. It is also crucial to analyze what kinds of mistakes the model makes and to ensure it performs fairly across different subgroups of data.

## A Toolkit of Algorithms

Different problems call for different algorithms, each with their own strengths. Some excel at recognizing patterns in images or text, while others are better suited for predicting numerical values or uncovering hidden structures in data. The right choice depends on the nature of the problem, the type of data available, and the desired balance between accuracy, speed, and interpretability.

The most common families of algorithms are described below:

- **Linear Models:** Are simple and interpretable. They learn to assign a weight to each input feature, showing how each variable contributes to the outcome. Linear Regression (used for predicting continuous values) and Logistic Regression (used for classification) are foundational examples.
- **Decision Trees:** Represent decisions through a tree-like sequence of questions (e.g., "Is age greater than 30?"). They are easy to interpret but can become overly tailored to the training data, a problem known as overfitting.
- **Ensemble Methods:** These combine multiple simple models (like decision trees) to create a more powerful and robust one; Random Forests build many trees in parallel and average their results; Gradient Boosting (e.g., XGBoost) builds trees sequentially, with each new tree learning to correct the errors of the previous ones. These are often the top performers on structured data.

- **Support Vector Machines (SVMs):** Are effective for separating data into categories, even when the boundary between them is complex or curved.
- **K-Nearest Neighbors (KNN):** Is a simple algorithm that makes predictions by looking at the most similar examples in the data. For example, it classifies a new item based on the majority of its closest neighbors.
- **Neural Networks:** Are highly flexible models composed of interconnected layers. They excel at automatically learning hierarchical features from complex data such as images, sound, and text. Specialized architectures include Convolutional Neural Networks (CNNs) for image and spatial data, Recurrent Neural Networks (RNNs) for sequential data such as time-series, text, and speech, and Transformers for handling long-range dependencies in sequential data like language and audio.

## Where It Is Used: Applications

Supervised learning is remarkably versatile because it can learn from examples to make accurate predictions across many domains:

**Computer Vision**

Image classification (e.g., identifying a cat in a photo), object detection (e.g., locating pedestrians in a self-driving car›s view), and medical image analysis.

**Natural Language Processing (NLP)**

Spam detection, sentiment analysis (e.g., determining if a review is positive or negative), machine translation, and named entity recognition.

**Finance and Commerce**

Credit scoring, fraud detection, and forecasting sales or stock prices.

**Healthcare**

Assisting in disease diagnosis from symptoms or medical scans, and predicting patient outcomes.

### Advantages and Limitations

Supervised learning performs exceptionally well when labeled data are abundant and accurately reflect real-world conditions. However, labeling data can be time-consuming and costly, and models may degrade if future data differ significantly from the training examples (a problem known as distribution shift).

## 3.3 Unsupervised Learning Explained

While supervised learning uses labeled data with clear answers, unsupervised learning, by contrast, seeks to uncover hidden structures and patterns in data that has no labels or predefined outcomes. It is like being handed a vast library of untitled books and organizing them by topic, author, or writing style. In the same way, unsupervised learning explores data independently to discover its underlying organization.

### The Core Idea: Learning from Data's Natural Structure

The key characteristic is the absence of target labels. The algorithm receives only raw input data and must find meaningful patterns without any guidance on what to look for. This makes it powerful for exploratory analysis and for situations where labeling data is impractical, but vast amounts of raw data are available.

## Common Tasks and Goals

Unsupervised learning plays a vital role in exploring and understanding data when no predefined labels are available. It is primarily used to uncover natural groupings, simplify complex datasets, and detect unusual patterns that stand out from the norm. Its main objectives include:

- **Clustering:** Automatically grouping similar data points together, such as segmenting customers for targeted marketing without pre-defined categories.
- **Dimensionality Reduction:** Simplifies data by reducing large, complicated datasets with many variables into a simpler form so patterns become easier to see and understand.
- **Anomaly Detection:** Identifying rare or unusual items that deviate from normal patterns, crucial for detecting fraudulent transactions or system failures.
- **Feature Learning:** Discovering useful new representations of the data, which can then be used to improve the performance of other machine learning models.

## A Toolkit of Algorithms

A wide range of algorithms exists within unsupervised learning, each designed to uncover specific kinds of patterns or structures hidden in data. These algorithms help reveal relationships, simplify complex datasets, and expose underlying trends that might otherwise go unnoticed.

Some of the most commonly used algorithms include:

- **K-Means Clustering:** A popular method that partitions data into a pre-specified number of groups (k), where each point belongs to the group with the nearest center. It is efficient but requires estimating the number of groups in advance.
- **Hierarchical Clustering:** Builds a tree-like model of clusters (a "dendrogram"), revealing the data's natural groupings at different levels of detail, from very broad to highly specific.
- **DBSCAN:** A clustering method that identifies groups based on the density of data points. It excels at finding irregularly shaped clusters and can automatically detect points that are outliers or "noise."
- **Principal Component Analysis (PCA):** The most common technique for dimensionality reduction. It identifies the most meaningful "directions" to view the data from, compressing it while preserving as much of the original information as possible.
- **Autoencoders:** Neural networks trained to compress data into a simplified code and then reconstruct the original data from that code. The learned compressed representation serves as a powerful, compact summary of the input.

## Where It Is Used: Applications

Unsupervised learning helps uncover hidden structure within data, organizing what was unorganized and finding meaningful signals in the noise. It is essential for making sense of the world's vast amounts of unlabeled information across many fields, including:

**Customer Segmentation**

Grouping users or customers for targeted advertising and services.

**Scientific Discovery**

Finding new types of galaxies or grouping genes with similar activity.

**Data Preprocessing**

Cleaning and simplifying data to improve the performance of supervised learning models.

**Recommendation Systems**

Grouping similar products or content to make better, more relevant recommendations.

**Anomaly Detection**

Used in cybersecurity, fraud detection, and industrial monitoring to flag unusual behavior or potential failures.

### Advantages and Limitations

Unsupervised learning is valuable when labeling is impractical or impossible. It supports exploratory analysis and reveals insights that humans might miss. However, it can be ambiguous—there is no clear "right answer," and results must often be interpreted carefully within domain context.

## 3.4 Reinforcement Learning Explained

Reinforcement Learning introduces a new dimension—learning through experience and feedback from actions. It centers on a key question: *How can an intelligent agent learn to make the best possible decisions in a dynamic, interactive environment?* Unlike other machine learning approaches that train on fixed datasets, reinforcement learning learns by doing. The system interacts with its environment, observes the results of its actions, and continuously adjusts its strategy to maximize long-term rewards.

## The Core Framework: Agents, Environments, and the Feedback Loop

The entire process is based on the interaction between an agent (the decision-maker) and an environment (everything it interacts with). This interaction happens repeatedly in a cycle that defines how reinforcement learning works:

- **Observe the Situation:** The agent looks at its current surroundings or condition. For a robot, this might include its camera view or sensor readings; in a game, it could be the state of the board.
- **Choose an Action:** Based on what it has learned so far, the agent decides what to do next.
- **Get Feedback:** The environment reacts to that action and gives the agent a reward or penalty that indicates how good or bad the decision was.
- **Learn and Repeat:** Using this experience, the agent updates its strategy to make better choices in the future.

## Key Concepts and The Learning Challenge

Several key ideas form the foundation for understanding how reinforcement learning works and why it can be so challenging. These concepts explain how an agent makes decisions, learns from experience, and balances short-term rewards with long-term goals:

- **The Policy:** This is the agent's brain, a function that maps states to actions. It can be a simple lookup table or a complex neural network. The entire goal of reinforcement learning is to find the optimal policy.

- **The Reward Hypothesis:** The agent›s sole objective is to maximize the sum of all future rewards, often with a discount factor that makes immediate rewards slightly more valuable than distant ones. This simple goal can lead to highly complex behaviors.
- **The Exploration vs. Exploitation Trade-off:** This is the fundamental dilemma. Should the agent exploit its current best-known actions to reap rewards, or should it explore new, uncertain actions that might lead to even greater long-term payoff? Balancing this is key to success.
- **The Credit Assignment Problem:** In a long sequence of actions, how does the agent determine which specific actions were responsible for a final outcome (like winning or losing a game)? Reinforcement learning algorithms must solve this to learn effectively.

## A Toolkit of Advanced Algorithms

Reinforcement learning researchers have developed several families of algorithms to solve these challenges. Each approach represents a different way of learning from experience–some focus on directly improving decision strategies, while others learn how to predict long-term rewards more accurately. Together, they form the foundation of modern reinforcement learning applications, from robotics to game-playing AI. The most widely used categories of reinforcement learning algorithms are summarized below:

- **Value-Based Methods (Learning Which Actions Work Best):** These algorithms, such as the foundational Q-Learning, focus on learning which actions are most rewarding in different situations. The idea is simple: the system keeps track of how good each possible action is and learns, through trial and error, which ones lead to better outcomes. Once it understands which actions bring the highest rewards, it can consistently make the best choices. Deep Q-Networks (DQNs) extend this idea by using neural networks to handle complex environments, such as learning Atari games from raw pixels.

- **Policy-Based Methods (Directly Optimizing Behavior):** Instead of learning values and deriving a policy, these methods directly parameterize and optimize the policy itself. Algorithms like REINFORCE adjust the policy parameters in the direction that increases the probability of taking actions that led to high rewards. This approach is naturally suited for environments with continuous or complex action spaces.
- **Actor-Critic Methods (The Best of Both Worlds):** This hybrid architecture is the backbone of most modern reinforcement learning successes. It uses two components: an Actor (the policy that selects actions) and a Critic (a value function that evaluates those actions). The Critic's feedback allows the Actor to update its policy much more efficiently and stably than it could alone. State-of-the-art algorithms like Proximal Policy Optimization (PPO) and Soft Actor-Critic (SAC) are Actor-Critic methods.
- **Model-Based Methods (Learning to Simulate and Plan):** These algorithms take a different approach by first learning a model of the environment, a predictive simulation of how the environment will respond to actions. With a good model, the agent can plan ahead by "thinking" about potential future sequences of actions without interacting with the real environment, leading to superior sample efficiency. This is a key component of systems like AlphaZero, which combines a learned model (used in Monte Carlo Tree Search) with policy and value networks.

## Where It Is Used: Pushing the Boundaries of AI

Reinforcement learning is the go-to approach for problems that involve making a sequence of decisions and planning for long-term success. It shines in environments where each action affects what happens next, allowing systems to learn effective strategies through trial, feedback, and continuous improvement. Its impact spans many fields:

- **Game Playing:** Reinforcement learning has made history, with AlphaGo and AlphaStar beating world champions in Go and StarCraft

II. These systems discovered non-human strategies through self-play, demonstrating reinforcement learning's creative potential.

- **Robotics:** Reinforced learning enables robots to learn complex dexterous skills–from walking and running to manipulating objects with fine motor control–directly from practice, often by training first in photorealistic simulators.
- **Autonomous Systems:** Self-driving cars use reinforced learning for high-level decision-making like lane merging and navigation. It is also used to manage complex infrastructure, such as optimizing energy usage in data centers, resulting in significant cost and energy savings.
- **Business and Finance:** Reinforced learning powers recommendation systems that optimize for long-term user engagement over time. In finance, it is explored for algorithmic trading strategies and portfolio management.
- **Other Frontiers:** Reinforcement learning is being used to optimize chemical reactions in laboratories, design more efficient computer chips, and guide personalized treatment plans in healthcare. One especially important development is Reinforcement Learning from Human Feedback (RLHF), a method that helps train large language models such as ChatGPT, to better align their responses with human preferences and values.

## Advantages and Limitations

Reinforcement learning excels in sequential decision-making problems where actions influence future outcomes. However, it requires extensive experimentation and significant computational resources. Designing appropriate reward functions and ensuring safety during learning remain major challenges. Despite these difficulties, its ability to discover complex, adaptive behaviors makes it one of the most powerful–and rapidly advancing–approaches in modern AI.

## 3.5 Comparing Machine Learning Approaches

Each learning paradigm—supervised, unsupervised, and reinforcement learning—approaches the challenge of acquiring knowledge in a distinct yet complementary way. They differ in the type of feedback they receive, the goals they pursue, and the nature of their learning process. Understanding these distinctions clarifies why different algorithms are suited to different categories of problems.

Supervised learning is guided by direct instruction. The system learns from labeled examples in which the correct output is already known. Its objective is to predict the right outcome for new inputs by identifying consistent relationships between features and labels. This paradigm is highly effective when large, high-quality datasets exist, such as medical image classification, loan approval prediction, or email spam detection. Algorithms like linear regression, decision trees, and neural networks are commonly used to fit these mappings between inputs and outputs.

Unsupervised learning, by contrast, operates without a teacher. It receives only raw, unlabeled data and must uncover patterns, clusters, or hidden structures on its own. Rather than predicting specific outcomes, it seeks to organize information in meaningful ways, revealing relationships that are not immediately obvious. This paradigm is essential for exploratory data analysis and applications like customer segmentation, topic discovery in text data, and anomaly detection. Common techniques include methods for grouping similar items (like k-Means), reducing complex data into simpler forms (such as PCA), and visualizing patterns using tools like t-SNE, UMAP, or autoencoders.

Reinforcement learning takes yet another path. Instead of learning from static data, it learns through interaction. The system, known as an agent, explores an environment, takes actions, and receives feedback in the form of rewards or penalties. Over time, it refines a policy, a strategy for choosing actions that maximize cumulative reward. This approach excels in sequential decision-making tasks where actions influence future outcomes, such as game playing, robotics, and real-time resource management. Algorithms such as *Q-Learning*, *Policy Gradient*, and *Actor–Critic* models balance the need to explore new strategies with the need to exploit what has already been learned.

While supervised learning focuses on fitting known examples, unsupervised learning emphasizes discovering structure, and reinforcement learning centers on acting and improving through experience.

Each paradigm has its strengths and limitations: supervised methods depend on the availability of labeled data, unsupervised methods require interpretive insight to make sense of discovered patterns, and reinforcement methods demand extensive experimentation and well-designed reward systems.

In practice, many modern AI systems combine aspects of all three. For example, a self-driving car uses supervised learning for visual recognition, unsupervised learning to detect novel traffic scenarios, and reinforcement learning to optimize driving strategies. Understanding these paradigms as complementary rather than competing perspectives provides a more holistic grasp of how intelligent systems learn, adapt, and improve over time.

## 3.6 Hybrid and Emerging Machine Learning Approaches

In practice, the boundaries between learning paradigms are increasingly blurred. Modern AI systems often combine multiple approaches to enhance efficiency, flexibility, and adaptability. As machine learning advances, these hybrid methods merge the strengths of different learning styles to overcome limitations of any single paradigm. Notable examples include:

- **Semi-Supervised Learning:** Uses a small amount of labeled data alongside large volumes of unlabeled data to improve learning efficiency.
- **Self-Supervised Learning:** Generates its own training labels from data, a key technique behind large language models and foundation models.

- **Transfer Learning:** Reuses knowledge gained from one task or domain to accelerate learning in another.
- **Federated Learning:** Trains models across distributed devices while preserving data privacy.

These hybrid and self-supervised approaches have driven the emergence of large language models and multimodal systems, bridging the gap between narrow task performance and broader forms of general intelligence. Together, they demonstrate how machine learning continues to evolve, adapting to data availability, computational constraints, and growing ethical considerations.

## 3.7 Key Takeaways

Machine learning has fundamentally redefined artificial intelligence, shifting it from a paradigm of explicit programming to one where systems learn directly from data. This transformation, which is driven by three core learning paradigms, supervised, unsupervised, and reinforcement learning, underpins the intelligent technologies that shape modern life, from recommendation systems to medical diagnostics and autonomous systems.

Each paradigm offers a distinct way for machines to acquire intelligence:

- **Supervised learning** mirrors education with a teacher, learning from labeled examples to make precise predictions.
- **Unsupervised learning** resembles self-discovery, uncovering hidden structures and relationships in unlabeled data.
- **Reinforcement learning** reflects learning through experience, using feedback and rewards to master complex, sequential decision-making.

Together, these paradigms form a versatile toolkit for perception, reasoning, and action, the foundation of modern AI.

As the field evolves, hybrid approaches such as self-supervised, semi-supervised, and transfer learning are expanding what machines can learn and how efficiently they can do it. The next frontier lies in developing systems that are more data-efficient, interpretable, and trustworthy, capable of learning safely and adapting continuously in dynamic environments.

Current research is focused on several key challenges:

**Sample Efficiency**

Enabling systems to learn effectively from limited data, closer to human learning.

**Robustness and Safety**

Making models dependable and resistant to manipulation.

**Fairness and Interpretability**

Making AI decisions transparent, understandable, and free from harmful bias.

Emerging directions include meta-learning, where systems learn to improve their own learning process; causal reasoning, which seeks to understand not just correlations but true cause-and-effect relationships; and lifelong learning, where AI continues to evolve and adapt without forgetting prior knowledge.

Understanding these fundamentals–the paradigms, their strengths, and their limitations–provides the conceptual foundation for exploring how learning systems are designed, implemented, and applied. It prepares us to harness the power of today's machine learning technologies responsibly and to participate in shaping their transformative future.

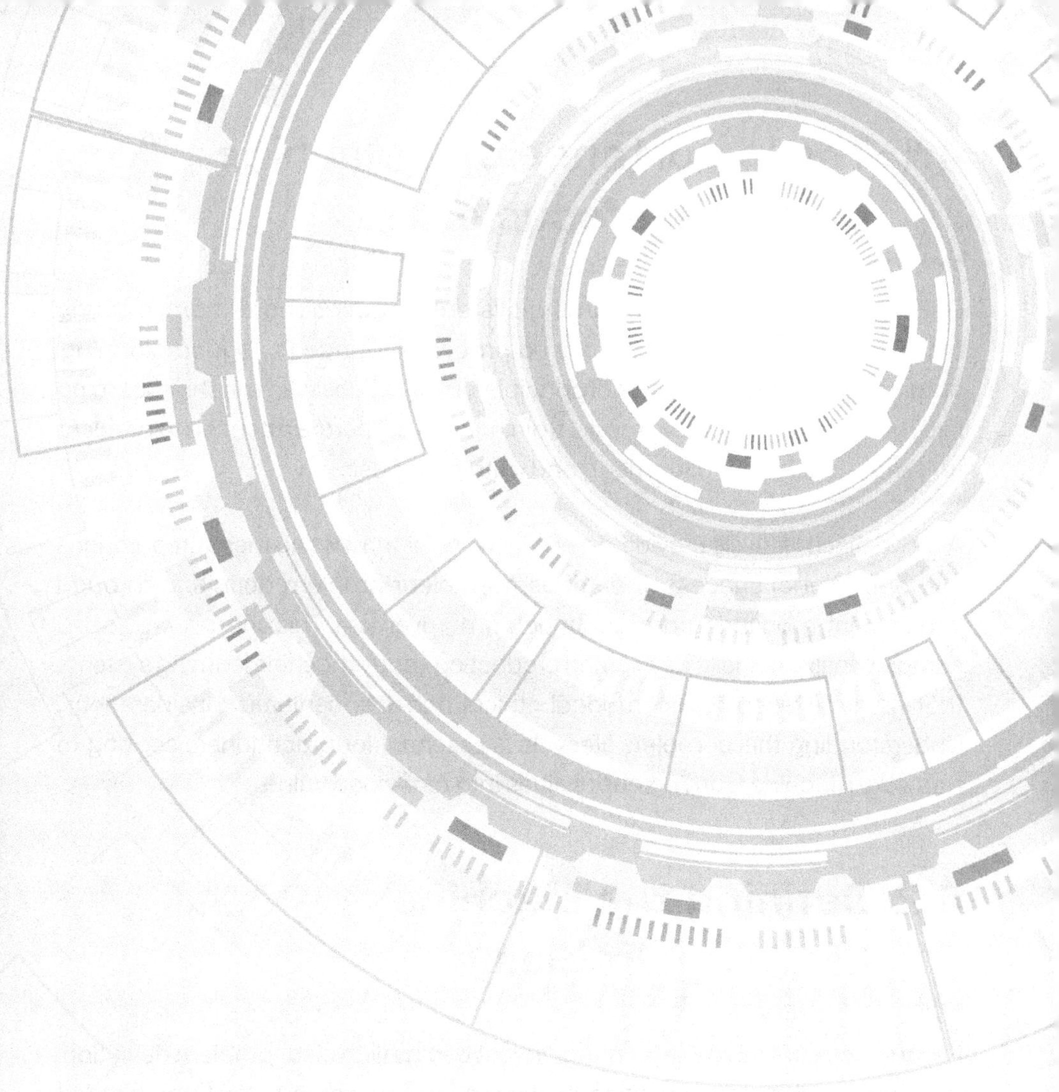

Chapter 4

# Planning and Data Foundations for ML

## 4.1 The Importance of a Machine Learning Lifecycle

Building effective machine learning systems requires more than selecting algorithms–it demands a structured process grounded in sound planning, data understanding, and ethical principles. Every successful machine learning initiative begins long before model training, with deliberate attention to problem framing, stakeholder alignment, and data readiness.

A well-defined lifecycle provides the framework for translating these preparations into operational success. It extends from clear problem definition through deployment, monitoring, and continuous improvement. This chapter provides a comprehensive guide to developing production-ready machine learning systems that integrate technical, operational, ethical, and user-centered considerations. Understanding this complete lifecycle is essential for practitioners seeking to translate machine learning capabilities into real-world value.

## 4.2 Defining the Machine Learning Problem

Every successful machine learning project begins with clear problem definition. Ambiguous or poorly specified problems lead to wasted effort, misaligned solutions, and systems that fail to deliver value despite technical sophistication.

Problem formulation translates business objectives or user needs into specific machine learning tasks. This requires understanding what outcome is desired, what constitutes success, and whether machine learning is appropriate for the problem.

## Identifying the Core Objective

Defines precisely what the system is meant to achieve and why it matters. Is the goal to predict which customers may leave, detect fraud, recommend products, or help diagnose diseases? A well-formulated objective not only directs technical development but also ensures that the machine learning effort supports measurable business or societal value. Linking the model's purpose to real-world impact keeps the team focused and prevents work on solutions that look impressive but don't matter.

## Determining Task Type

Defines what kind of problem the system must solve within the broader landscape of machine learning. Classification assigns inputs to distinct categories (e.g., spam vs. not spam), while regression predicts continuous values such as sales or temperature. Clustering uncovers natural groupings in data, reinforcement learning optimizes sequential decisions through feedback, and anomaly detection highlights deviations from expected patterns. Identifying the correct task type is critical–it determines which algorithms are suitable, what data is needed, and how success will be measured.

## Assessing Machine Learning Suitability

Determines whether machine learning is the right solution for the problem at hand. Machine learning is most effective when patterns exist in the data that are too complex to express through fixed rules, when sufficient high-quality examples are available, and when the system can tolerate some level of error. It is less suitable when data is limited, the task demands perfect accuracy, or simple rules can already solve the problem well. A thoughtful assessment at this stage prevents wasted effort and ensures that machine learning is applied where it can truly add value.

# 4.3 Defining Success Metrics and Constraints

Defining success metrics provides quantitative goals and enables objective evaluation. Metrics must reflect actual business value, not just technical performance. Clearly defined metrics guide both model development and decision-making by linking technical outcomes to organizational objectives. They also provide a benchmark for assessing progress, ensuring that the system's improvements translate into tangible real-world impact.

## Primary Metrics

Primary metrics measure how well the model meets its main goal. In classification, this may be accuracy, precision, recall, or F1–depending on whether false positives or false negatives matter more. Regression employs mean squared error, mean absolute error, or R-squared. Recommendation systems track click-through rates, conversion rates, or user engagement. The selection of the primary metric directly guides how a model is developed.

## Secondary Metrics

Secondary metrics capture additional considerations. Beyond accuracy, systems may need to respond quickly, handle large volumes of requests, or perform fairly across different demographic groups. Memory and computing limits can also restrict how complex a model can be.

## Constraints and Requirements

Constraints and requirements define the practical boundaries within which a machine learning solution must function. Real-time systems need fast responses, and edge devices need small models that run well on limited hardware. Regulated industries emphasize transparency and interpretability, and privacy-sensitive applications must minimize or anonymize data use. Resource limits—such as computation, storage, or energy—further shape what is feasible. Clearly identifying these constraints early helps prevent wasted effort on solutions that cannot meet real-world operational needs.

## 4.4 Aligning Stakeholder Needs

Machine learning systems serve diverse stakeholders with potentially conflicting interests. Aligning expectations prevents disappointment and ensures solutions address actual needs. Early collaboration with business leaders, technical teams, and end-users helps clarify priorities, define success criteria, and manage trade-offs between accuracy, cost, and interpretability. Continuous communication throughout the project lifecycle strengthens trust and promotes shared ownership of outcomes.

### Identifying Stakeholders

This step maps out everyone who interacts with, is affected by, or has authority over the system. End users directly engage with predictions or recommendations, while domain experts contribute knowledge and validate outcomes. Business leaders define goals and allocate resources, and engineers design, build, and maintain the system. Some stakeholders, such as customers or communities, may not use the system directly but still experience its outcomes or decisions.

### Understanding Requirements

This step focuses on identifying and reconciling the priorities of different stakeholders. Users often value ease of use, speed, and reliability. Domain experts may emphasize interpretability and alignment with professional standards. Business leaders prioritize ROI and strategic impact, while engineers seek maintainable and scalable architectures. Recognizing these diverse needs early ensures that the final system balances usability, trustworthiness, and business value.

### Managing Expectations

Managing expectations means clearly communicating both the potential and the limitations of machine learning. These systems produce probabilities rather than definite answers, and their performance depends on good data and careful development. Even well-designed models may struggle with rare or unforeseen scenarios not captured in training data, and performance can drift as real-world conditions evolve. Establishing realistic expectations early fosters trust, reduces disappointment, and helps stakeholders understand that responsible AI aims for accuracy, reliability, and ethical balance—rather than flawless perfection.

## 4.5 Analyzing Data Requirements

Data forms the foundation of machine learning systems. The quantity, quality, and relevance of data fundamentally constrain what models can learn and how well they generalize. Understanding data requirements guides collection efforts and identifies potential gaps. A clear data strategy ensures that the right information is gathered efficiently, reducing rework and enabling more reliable model development.

### Determining Data Needs

This step leads to a specification of what information is necessary for the task. Supervised learning requires labeled examples, inputs paired with correct outputs. The number of examples needed depends on problem complexity, input dimensionality, and desired performance. Simple problems with low-dimensional inputs may need hundreds of examples; complex problems with high-dimensional inputs may require millions.

## Identifying Data Sources

This step involves mapping where all relevant data comes from. Internal sources may include databases, logs, sensors, or transaction records. External sources can include public datasets, purchased data, or connections to other systems through APIs – tools that allow different software applications to share information automatically. Data partnerships may also provide access to proprietary datasets, while user-generated content creates new data through everyday interactions with digital platforms. Each source differs in availability, cost, quality, and legal constraints.

## Assessing Data Feasibility

This step involves evaluating whether sufficient data can be obtained. Some problems lack adequate data due to rarity (rare diseases), privacy (medical records), expense (labeled satellite imagery), or novelty (new products). Data feasibility assessment prevents pursuing projects doomed by fundamental data limitations.

## Data Balance and Representativeness

Ensuring that the dataset is balanced and representative of the real-world population is critical for building fair and reliable models. Imbalanced data–where some categories, outcomes, or demographic groups are underrepresented–can cause bias and lead to poor generalization. An effective analysis examines whether all relevant segments are adequately captured, and, if not, whether techniques such as resampling, data augmentation, or synthetic data generation can help achieve greater balance. Addressing these issues early improves both the model's performance and its fairness across diverse users.

## 4.6 Strategies for Data Acquisition

Data acquisition implements collection strategies, balancing coverage, quality, cost, and timeliness. Effective acquisition involves selecting appropriate sources, ensuring consistent data formats, and maintaining ethical and legal compliance throughout the process. Establishing automated pipelines and validation checks helps sustain data reliability and scalability as the system evolves.

### Automated Collection

Automated data collection uses software systems to gather information at scale and speed. Sensors continuously record environmental or operational measurements, logging systems track user interactions, APIs retrieve data from external platforms, and web scrapers collect data from online sources. This approach enables continuous, large-scale data acquisition but also demands vigilance—automated pipelines can easily introduce systematic biases, redundant entries, or gaps if not carefully monitored and validated.

### Manual Collection

Manual collection plays a critical role in ensuring data quality and context. It often relies on three key sources: crowdsourcing platforms that distribute labeling tasks across many contributors for scalability, domain experts who apply specialized knowledge for accuracy in complex tasks, and internal teams who generate proprietary datasets aligned with organizational needs. Though time-intensive, manual efforts often yield higher accuracy and interpretability—qualities essential for reliable model performance.

### Existing Datasets

Leveraging existing datasets can dramatically accelerate development by providing ready-made, well-structured data resources. Academic benchmarks allow teams to measure progress against established standards, while pre-trained models enable knowledge transfer from large-scale datasets to new applications with less data. In addition, synthetic data, created through simulations or generative models, can expand small datasets, adding variety while reducing collection costs. However, synthetic data must still be validated to ensure it reflects real-world distributions and avoids bias drift.

## 4.7 Data Ethics and Legal Rights

Data collection must respect legal rights, ethical principles, and social norms. This entails obtaining informed consent, ensuring transparency in how data is used, and complying with regulations like GDPR. Furthermore, it requires a proactive commitment to fairness, avoiding the reinforcement of biases that could lead to discriminatory outcomes.

### Legal Compliance

Legal compliance ensures that all data collection and usage activities adhere to relevant laws and regulations. Privacy frameworks such as GDPR and CCPA impose strict limits on how personal data can be gathered, stored, and shared. Terms of service govern what may be accessed through APIs or web scraping, while copyright laws protect creative and intellectual works. Data licensing agreements further define permitted uses and redistribution rights. Failure to comply with these requirements can lead to serious legal consequences, financial penalties, and lasting reputational harm.

### Informed Consent

Informed consent requires clearly communicating with individuals about what data is being collected, why it is needed, how it will be used, and who will have access to it. Transparency about data retention and sharing practices upholds individual autonomy and helps build trust. Opt-in consent, where users actively agree before data is gathered, is generally stronger and more ethical than opt-out mechanisms. Additional safeguards are essential when collecting data from children, vulnerable populations, or involving sensitive personal information.

### Ethical Considerations

Ethical responsibility extends beyond compliance with laws. Gathering data without a clear, justified purpose risks exploitation and erosion of public trust. The surveillance capabilities enabled by machine learning amplify concerns about privacy, autonomy, and potential misuse. Collecting data from marginalized or at-risk groups requires special care to avoid reinforcing inequities or causing harm. Responsible data practices should always weigh potential benefits against possible risks, ensuring that data serves human well-being rather than undermining it.

### Data Anonymization and Privacy Protection

Protecting individual privacy is central to ethical data management. Organizations often remove or mask personal details before analysis to prevent identification. Advanced techniques such as differential privacy introduce controlled statistical noise, safeguarding individual records while maintaining analytical value. When applied thoughtfully, these methods help organizations comply with privacy regulations, reduce risk, and maintain public trust in data-driven innovation.

## 4.8 Preparing Data for Machine Learning

Raw data is seldom ready for immediate use in model training. Through preprocessing, it must be transformed into a structured and reliable form that supports effective learning. This process includes cleaning to detect and correct errors, resolve inconsistencies, and remove anomalies that could mislead the model. Well-prepared data ensures that learning is based on accurate, consistent, and representative information, establishing a solid foundation for credible, trustworthy results.

## Handling Missing Values

This task is designed to identify, manage, and resolve issues of incomplete data to ensure the integrity and usability of the dataset for downstream analysis. Deletion removes examples or features with missing values, reducing dataset size but ensuring completeness. Imputation fills missing values with estimates–means, medians, modes, or model predictions. Indicator variables flag missing values, allowing models to learn patterns associated with missing data. The appropriate approach depends on why data is missing and how much is absent.

## Identifying Outliers

This task detects anomalous values by statistically analyzing patterns and identifying significant deviations from the expected range, which is crucial for ensuring data quality. This is typically achieved through several complementary approaches: statistical methods flag values that fall far from the average; domain knowledge identifies physically impossible or logically inconsistent values; and visualization techniques reveal unusual patterns. Outliers may represent errors requiring correction, or genuine extreme values requiring special handling to prevent them from dominating model behavior.

## Correcting Errors

The error correction phase systematically resolves data quality issues uncovered during profiling and validation to create a clean and reliable dataset. This involves correcting typographical mistakes, merging or removing duplicate records, and standardizing inconsistent formats such as dates, units, or categorical labels. Referential integrity checks ensure that related records connect correctly across tables or systems. While automated processes efficiently handle recurring or rule-based errors, manual review remains essential for resolving ambiguous or context-dependent cases.

## 4.9 Transforming Data for Modeling

Data transformation is a critical preprocessing step that converts raw, often unstructured features into refined, structured representations that are numerically and semantically suitable for the specific requirements of machine learning algorithms, ultimately enhancing their ability to detect patterns and learn effectively. It bridges the gap between real-world data complexity and the mathematical precision required by models, ensuring that inputs are both meaningful and computationally efficient.

### Normalization and Scaling

This process standardizes the numerical range of features to ensure they contribute equally to model learning. For example, when using a person's age (typically under 100) and annual income (often in the tens of thousands) as inputs, a model would inherently assign greater importance to income due to its larger numerical magnitude, regardless of its actual predictive value. Scaling resolves this by transforming all features to a common scale, such as a 0 to 1 range, preventing any single variable from dominating the analysis due to its scale alone. Furthermore, certain scaling methods, such as those based on robust statistics, are designed to mitigate the influence of outliers, enhancing the stability and reliability of the model.

### Encoding Categorical Variables

Machine learning models require numerical input, which necessitates converting text-based categories into a numerical format through a process known as encoding. The simplest method, label encoding, assigns a unique integer to each category (e.g., "red"=1, "blue"=2, "green"=3). However, a significant drawback is that the model might mistakenly assume a non-existent order, for example, thinking that "green" is better or more than "red" simply because 3 is larger than 1. A more common and reliable method is one-hot encoding, which creates separate yes/no columns for each category so the model treats them as independent. The choice of encoding strategy is critical, as it fundamentally shapes the model's understanding of categorical data.

### Feature Engineering

Feature engineering is the process of creating new, more informative input variables from existing raw data to enhance a model's predictive capability. Rather than relying solely on base data, this technique involves constructing derived features that provide clearer signals to the learning algorithm. For example, from the raw length and width of a house, one can engineer a new feature representing total square footage. By applying domain knowledge to combine or transform existing variables—such as calculating financial ratios from balance sheet figures or extracting meaningful statistics from text—this process can significantly improve model performance. Although advanced methods like deep learning can automate some feature discovery, deliberate feature engineering remains a powerful and often essential step for building effective models.

## 4.10 Splitting Data for Training and Testing

Data splitting strategically partitions the available dataset into distinct subsets for training, validation, and testing. This practice is essential for fine-tuning the model and, most critically, providing an unbiased estimate of performance on unseen data to prevent overfitting. Properly designed splits ensure that evaluation results reflect real-world conditions and that the model's performance generalizes beyond the training environment.

## Training Set

The training set is the subset of data used to optimize a model's internal parameters through the learning algorithm. It must be sufficiently large to capture the underlying patterns in the data, yet balanced against the need to reserve adequate data for unbiased evaluation in the validation and test sets. Typical allocations for the training set range from 60% to 80% of the total available data, a proportion that depends heavily on the overall dataset size.

## Validation Set

The validation set is used to tune hyperparameters and guide the selection of the best-performing model. It provides unbiased performance estimates during the development cycle without contaminating the final test set. It is critical that the validation data accurately represents the real-world data distribution the model will encounter upon deployment, with typical allocations ranging from 10% to 20% of the total dataset. If the validation set doesn't reflect real data, the model will be tuned incorrectly and won't generalize well.

## Test Set

The test set provides the final, unbiased evaluation of a model's performance after development is complete. It must remain completely isolated from the training and validation processes to prevent any information leakage through preprocessing, feature selection, or other development decisions. Because it is held out from all aspects of model creation, its performance is the best estimate of real-world performance, with typical allocations ranging from 10% to 20% of the total data. Consequently, this final metric is the most reliable indicator of how the model will generalize to new, unseen data.

## Stratification

Stratification ensures that split datasets maintain the original proportions of classes or other important characteristics. This prevents splits where some categories appear too often in one subset and not at all in another, which can distort training and evaluation. For instance, temporal splits respect time ordering in time-series data to prevent information leakage from the future to the past. Without stratification, a model trained on an unrepresentative split may perform poorly on minority classes or fail to generalize effectively.

## 4.11 Expanding Training Data with Augmentation

Data augmentation expands the training dataset by creating modified versions of existing samples. It helps improve model generalization, particularly when available data is limited or imbalanced.

Traditional augmentation methods–like rotating, flipping, cropping images, or adding noise–create simple variations of existing examples. Modern synthetic data generation, by contrast, uses advanced techniques like GANs and diffusion models to create entirely new, realistic samples. These synthetic approaches are especially valuable when real data is limited, sensitive, or difficult to collect. However, the quality of synthetic data must always be validated against real-world distributions to prevent bias drift and ensure realistic model behavior.

### Image Augmentation

Image augmentation enhances robustness by applying transformations such as rotation, flipping, or color adjustment. These altered images retain their original labels, teaching the model that a cat is still a cat whether it appears upside down or in different lighting. This helps the model focus on essential object features rather than superficial variations.

## Text Augmentation

Text augmentation creates new training examples by rewording existing sentences, such as swapping synonyms or rephrasing passages, without changing meaning. This provides a richer and more diverse language set, helping the model understand intent rather than memorizing specific phrasing.

## Audio Augmentation

Audio augmentation applies similar principles to sound data by adjusting speed, pitch, or background noise. These modified clips retain their original labels, training the model to recognize words or sounds correctly even under varying audio conditions. The crucial rule across all types of augmentation is that transformations must never alter the core meaning, as doing so confuses the model and harms learning.

## Data Versioning and Reproducibility

As datasets evolve through processes like cleaning, transformation, and augmentation, tracking changes becomes essential for reproducibility and accountability. Modern tools like DVC (Data Version Control) and MLflow Data Registry let teams version datasets, track experiments, and link each model to the exact data used to train it. Incorporating data version control strengthens transparency, supports collaboration, and enhances the integrity of machine learning pipelines.

Chapter 5

# Modeling, Evaluation, and Deployment of Machine Learning Systems

# 5.1 Building and Training Machine Learning Models

Building an effective machine learning system requires more than just developing a model. It involves a disciplined process of model design, training, evaluation, validation, and deployment, with each stage contributing to the reliability, performance, and ethical soundness of the system. This chapter explores that journey, bridging the gap between experimentation and production-grade AI.

This process begins with a foundational choice: selecting the appropriate learning algorithm to form the core of the system. The selected algorithm determines how the model interprets data, learns patterns, and adapts over time. Choosing wisely ensures that the system aligns with both the complexity of the problem and the practical constraints of data availability, computation, and interpretability.

## Algorithm Families and Selection Principles

Choosing the right algorithm profoundly affects performance, interpretability, and development effort. Different algorithm families have unique characteristics suited to different data and goals. Understanding these distinctions helps practitioners match the strengths of each approach—such as simplicity, flexibility, or scalability—to the specific needs of a problem.

### Linear Models (The Straightforward Rulebook)

These models find simple, straight-line rules to make predictions. They are fast, easy to understand, and don't need much data. However, they cannot handle curved or non-linear patterns—situations where the relationship between inputs and outputs bends or changes direction—unless those patterns are added to the data beforehand.

## Tree-Based Models (The Step-by-Step Flowchart)

These models make decisions using a structure similar to a flowchart, asking a series of yes/no questions. They are powerful for finding complex, non-linear patterns and don't require much data cleanup. The downside is that they can become overly specific to the training data and don't always generalize well to new situations.

## Support Vector Machines (The Maximum Margin Separator)

Think of this model as trying to find the widest possible "street" to separate two groups of data points. It's effective for data with many features and can create complex boundaries using special tricks. Its limitations are that it can be slow with huge amounts of data and needs precise tuning to work well.

## Neural Networks (The Complex Brain)

Inspired by the human brain, neural networks are highly flexible models capable of learning complex patterns, making them state-of-the-art for tasks such as image recognition, language processing, and speech analysis. They include a range of architectures, such as feedforward networks for general tasks, CNNs for images, RNNs for sequences, transformers for long-range relationships, and diffusion models used in modern generative AI. The trade-off is that these models often act as "black boxes," which are difficult to interpret and require vast amounts of data and computing power to train effectively.

## Emerging Architecture

Recent advances have expanded the scope of deep learning beyond traditional architectures like CNNs and RNNs. Transformers now underpin most large-scale language and vision models due to their ability to handle long-range dependencies efficiently. Diffusion models have become foundational in generative AI, enabling the creation of high-quality images, audio, and video content. Meanwhile, foundation models—large, pre-trained systems that can be fine-tuned for diverse tasks—represent a new paradigm in machine learning, emphasizing scalability, transfer learning, and cross-domain adaptability.

### Ensemble Methods (The Wisdom of the Crowd)

This approach combines the predictions from multiple models to get a more accurate and reliable result. By averaging opinions or letting models vote, it often delivers superior performance. The cost is increased complexity, longer computation times, and a loss of simple interpretability that can be obtained from a single model.

## Algorithm Selection Criteria

Algorithm selection balances multiple considerations based on problem characteristics and operational requirements. Factors such as data size, feature types, interpretability needs, and computational resources all influence which algorithm will deliver the best balance between accuracy, efficiency, and maintainability.

### Performance Requirements

Defining precise performance requirements is a critical first step, as different applications demand distinct trade-offs between metrics such as recall and precision. For instance, medical diagnosis prioritizes high recall to avoid missing critical cases, while spam filtering emphasizes high precision to prevent blocking legitimate messages. These differences guide the selection of algorithms best suited to each performance profile.

### Data Characteristics

The characteristics and structure of the available dataset impose fundamental constraints on which algorithms are most suitable. Factors such as dataset size, dimensionality, class balance, and temporal dependencies each call for different modeling approaches – ranging from simple models for small datasets to specialized architectures designed for sequential data.

### Interpretability Needs

The required level of model interpretability is not a fixed standard but varies significantly depending on the application's context and consequences. In regulated industries or high-stakes scenarios, the ability to explain a prediction is paramount, often favoring simpler, more transparent models like linear models or decision trees instead of complex 'black box' systems.

### Computational Constraints

Practical computational limits, whether for training or inference, play a decisive role in narrowing the field of feasible algorithms. These constraints include the need for real-time prediction speeds, limitations on hardware memory and power, and strict training time budgets, which collectively favor efficient and compact models, especially for edge deployment.

### Development Resources

The scope and ambition of any machine learning project are ultimately constrained by the available development resources, including time, expertise, and tooling. While simple algorithms facilitate rapid prototyping and are less demanding, implementing and tuning complex state-of-the-art methods requires significant specialized knowledge and experimental effort.

## Establishing a Starting Point (Baselines)

The establishment of performance baselines constitutes a critical methodological step in the machine learning development lifecycle, providing essential benchmarks for the comparative assessment of proposed models.

### Simple Heuristics

Simple baseline models, based on straightforward statistical or logical rules, establish a benchmark for minimum acceptable performance. Common examples include predicting the most frequent class (for classification) or the average value (for regression), as well as using basic rule-based systems. Any model must clearly surpass these simple benchmarks to justify the additional cost and development effort.

### Domain-Specific Methods

In numerous specialized domains, robust non-machine learning methodologies exist, including expert systems, classical statistical models, or physics-based simulations. These approaches often encapsulate considerable prior knowledge and can provide surprisingly competitive benchmarks. A rigorous comparison against these established methods is necessary to validate whether a machine learning paradigm offers a tangible advantage over conventional, and often more interpretable, alternatives.

### Existing Solutions

Benchmarking against existing solutions–whether prior in-house models, commercial tools, or state-of-the-art research systems–provides essential context for evaluating progress. It places new developments within the broader ecosystem, helping determine whether they represent a genuine advancement or simply an incremental improvement over current capabilities.

## Model Training: Measuring Error with Loss Functions

Model training is the process of teaching an algorithm to make accurate predictions by repeatedly adjusting its internal settings. It improves over time by learning from mistakes, gradually turning a blank slate into a working system. To improve in this way, the model needs a method for measuring how wrong its predictions are. This measure, called a loss function, acts like a scoring system that guides the model toward better performance.

### Classification Losses

When a model is choosing between categories (e.g., "cat" or "dog"), the scoring system penalizes wrong guesses. Some scoring methods are very strict about the model's confidence, while others are specially designed to handle situations where there are more examples of one category than another.

### Regression Losses

When a model predicts a continuous value, such as a house price, the chosen scoring system quantifies the magnitude of its prediction errors. The selection of this metric is critical, as some functions heavily penalize large errors, others treat all errors uniformly, while balanced approaches offer a compromise between these two extremes.

### Custom Losses

Sometimes, a standard scoring system does not align with the real-world objective. In these cases, a custom scoring system can be created. For example, a design might prioritize the correct ranking of items over their exact predicted values, or it might combine multiple objectives to ensure the model's learning is precisely aligned with the application's specific requirements. Specialized losses, such as Huber for stability and focal loss for imbalanced data, help improve performance and fairness.

## Model Optimization and Convergence

The optimization process is the iterative method through which a model learns by systematically reducing its prediction errors. In essence, the model makes a prediction, measures how wrong it was, and then uses this feedback to update its parameters, repeating this cycle thousands of times until its performance converges on an optimal level.

### Gradient Descent (Core Algorithm)

This is the fundamental learning process. The algorithm determines the direction in which its parameters should change to most effectively reduce error–this direction is called the gradient. The size of each adjustment, controlled by the *learning rate*, is crucial: a rate that is too high leads to instability, while one that is too low makes training unnecessarily slow.

### Stochastic Gradient Descent (Sgd): Efficiency Through Sampling

This variation of gradient descent uses small, random subsets (mini-batches) of data to calculate the gradient rather than the entire dataset. This makes the process much faster and often helps the model generalize better to new data. Advanced versions incorporate *momentum* to smooth out updates, preventing oscillations and accelerating convergence.

### Adaptive Optimizers: Smarter, Automated Tuning

Adaptive optimizers like *Adam*, *RMSProp*, and *Adagrad* are smarter versions of basic training methods. They automatically adjust how big each learning step should be for different parts of the model, based on what has worked well in past steps. This makes training faster and more stable, especially when the data is messy or uneven. Although they usually require less manual tweaking, they may not always reach the absolute best result. In practice, they are often combined with other techniques to keep training both quick and reliable.

### Stability Challenges and Mitigation

Deep neural networks can sometimes become unstable during training. The learning signals may become too strong or fade, making it hard for the model to improve. Techniques like batch normalization, skip connections, and good initialization help keep training steady. Skip connections, popularized by architectures called ResNets (Residual Networks), allow information to bypass certain layers, making it easier for very deep networks to learn effectively.

### Convergence Monitoring: Determining the Stopping Point

During training, the model's progress is tracked to decide when it has learned enough without overdoing it. One common method is to watch how the error (or "loss") changes over time–if it stops improving, the model may be done learning. Another safeguard, called *early stopping*, automatically ends training when performance on unseen data starts to get worse, preventing the model from memorizing the training examples instead of learning real patterns.

# Controlling Complexity and Preventing Overfitting

Underfitting occurs when a model is too simple to capture patterns in data; overfitting arises when it memorizes noise. Balancing these is central to model robustness. Regularization techniques prevent overfitting by constraining model complexity or adding noise during training. This forces the model to learn simpler, more robust patterns from the data rather than memorizing it, which improves its ability to perform well on new, unseen information.

## L1 and L2 Regularization

These methods work by penalizing a model for becoming too complex. L2 regularization encourages all weights to be small, leading to a model that considers all inputs but does not rely too heavily on any single one. L1 regularization is more extreme, as it actively pushes many weights to zero, effectively performing feature selection by telling the model to ignore certain inputs entirely.

## Dropout

During training, this technique randomly 'drops out,' or temporarily turns off, a subset of neurons for each training example. In a neural network, a neuron is a simple computing unit that receives inputs, applies a small calculation, and passes the result forward, much like a tiny decision-maker. Turning some of these neurons off at random prevents the model from relying too heavily on any single neuron and encourages it to learn more robust, redundant patterns, making the overall network more resilient.

## Data Augmentation

This technique regularizes a model by artificially expanding the training data with slightly altered versions of existing examples. By exposing the model to a wider variety of data, it learns to focus on the essential patterns and becomes less likely to memorize the training set.

### Early Stopping

This is a straightforward way to prevent overfitting by monitoring the model's performance on a validation set during training. The process is halted as soon as this validation performance begins to degrade, stopping the model before it has a chance to start memorizing noise in the training data.

### Batch Normalization

This method helps models learn faster and more reliably by keeping the values flowing through each layer at a steady scale. It prevents the learning signals from becoming too strong or too weak, which can otherwise slow or disrupt training. It also adds a bit of helpful randomness that improves generalization. While batch normalization works well for image models, newer systems like transformers often use other techniques–such as Layer Normalization or Group Normalization–that perform better with different kinds of data.

## 5.2 Training and Evaluation

Effective training converts theoretical model potential into measurable performance, while evaluation ensures that this performance is both reliable and meaningful. Together, they form the feedback loop that drives model improvement, balancing learning efficiency with real-world applicability. The following sections outline key strategies for structuring this process.

### Training Strategies

Model training and evaluation transform theoretical capability into measurable performance. This section explores strategies for efficient learning and metrics for fair, thorough assessment. Effective training strategies balance computational efficiency with model robustness, ensuring the system learns accurately while remaining adaptable to new data.

### Batch Size Selection

This is the choice of how many data examples to process before updating the model. A larger batch size gives a more reliable, stable update but requires more memory and can sometimes lead to a model that does not generalize as well. A smaller batch size makes the model update more frequently and can help it explore the data better, but each update is based on a less representative sample and is less precise.

### Learning Rate Scheduling

Instead of using a fixed step size for learning, this technique adjusts it over time. This process can be visualized as taking large, confident steps to cover ground quickly, followed by smaller, more precise steps to fine-tune the final result. Common strategies include gradually lowering the *learning rate* over time, occasionally resetting it to help the model move past stuck points, or using a "warm-up" phase that begins with very small steps to stabilize the model before increasing them.

### Transfer Learning

Transfer learning is a methodology that initializes a model with parameters pre-trained on a large, general-purpose dataset. This approach leverages the foundational feature representations already learned by the pre-trained model. Rather than training from scratch, the model is adapted, or fine-tuned, for a specific downstream task, which significantly reduces the computational resources, training time, and volume of task-specific data required.

### Curriculum Learning

This strategy involves training a model the way humans learn, by starting with easy examples and gradually introducing more difficult ones. By mastering simple concepts first, the model builds a stronger foundation, which often leads to better final performance and faster learning when it eventually tackles the hard problems.

# Evaluation Metrics and Methodology

Model evaluation assesses how well-trained models perform, using appropriate metrics and rigorous methodology to produce trustworthy performance estimates. Evaluation metrics are the scores used to measure a model's performance, and different tasks require different ways of keeping the score.

Classification models use metrics like accuracy, precision, recall, and F1-score. Regression models rely on measures such as mean squared error (MSE), which indicates how far predictions are from the true values on average, and $R^2$, which shows how much of the variation in the data the model explains. For models that output probabilities, calibration metrics such as the Brier score or reliability curves help assess how well predicted probabilities match actual outcomes, providing a more complete picture of model quality.

## Classification Metrics

These measures evaluate how well a model makes predictions that fall into categories, such as "spam" or "not spam." *Accuracy* shows how many predictions are correct overall, but it can be misleading if one category appears far more often than others. *Precision* tells how many of the predicted positives were actually correct, while *recall* shows how many real positives the model successfully found. The *F1 score* combines both precision and recall into a single balanced number. Another measure, *AUC-ROC*, measures how consistently the model assigns higher scores to positives than to negatives. The better the separation, the higher the AUC.

## Regression Metrics

These measures apply when the model predicts numbers instead of categories, such as prices or temperatures. *Mean Squared Error (MSE)* emphasizes larger mistakes more strongly, while *Mean Absolute Error (MAE)* treats all mistakes equally. *Root Mean Squared Error (RMSE)* expresses the error in the same units as the data, making it easier to understand. *R-squared* shows how much of the variation in the data is explained by the model–a higher value means better performance.

### Ranking Metrics

These measures evaluate how well a system orders its results–important for search engines and recommendation systems. They check whether the most relevant or useful items appear near the top of the list. Common examples include *Mean Average Precision (MAP)*, which looks at how accurate the list is overall; *Normalized Discounted Cumulative Gain (NDCG)*, which gives higher credit when important items appear earlier; and *Mean Reciprocal Rank (MRR)*, which focuses on how soon the first correct result appears. Together, these metrics reflect how helpful the ranked list feels to users in real-world situations.

### Business Metrics

These metrics translate technical performance into tangible business impact, such as revenue increase, cost reduction, or customer retention rate. The alignment of technical and business metrics is essential to ensure that model optimization directly supports overarching organizational objectives.

### Ensuring Reproducibility in Evaluation

To trust a model's results, others must be able to repeat the same experiment and get similar outcomes. This means keeping the setup consistent–using the same data, software versions, and settings each time. Controlling random factors during training also helps. Reproducibility ensures the results are real and reliable, not just a lucky outcome, and allows fair comparisons between different models or studies.

## Understanding and Reducing Model Errors

Rigorous evaluation methodology produces reliable performance estimates. This reliability is fundamental, as it ensures that the reported metrics reflect the model's true capabilities and its expected behavior when deployed in a real-world environment.

### Holdout Validation

This fundamental approach involves evaluating a model on a dedicated test set that was completely withheld during the training process. This provides an unbiased performance estimate, assuming the test set accurately reflects real-world conditions. However, a single test set can only provide one snapshot of performance and may not fully capture its variability.

### Cross-Validation

This technique is particularly valuable with limited data. It works by repeatedly splitting the data into different training and test sets, typically in a cycle called k-folding where each data point gets a turn to be in the test set. The results from all cycles are averaged to produce a more robust and reliable performance estimate, though this process is computationally intensive.

### Stratified Sampling

This is a sampling technique used to ensure that every data split, whether for training or testing, maintains the same proportion of key characteristics (like class labels) as the full dataset. This prevents an uneven or unrepresentative split from distorting the model's performance.

### Temporal Validation

For time-series data, the split must preserve the natural order of events. The model is trained on older, historical data and then tested on newer data, mirroring how it would be used in the real world to make future predictions. Randomly mixing the data would break this logic—allowing the model to "see" the future and giving it an unrealistic advantage.

## Model Error Analysis

Moving beyond a single aggregate metric, error analysis involves a detailed inspection of mistakes to understand a model's specific weaknesses and guide improvements. This process systematically categorizes failures—such as identifying which data subtypes are frequently misclassified or under what conditions the model is most unreliable—to provide actionable insights for refinement.

### Confusion Analysis

In classification tasks, this involves examining a confusion matrix to see which specific classes are most frequently mistaken for one another. Systematic confusion patterns can reveal ambiguous categories, a lack of informative features, or issues related to class imbalance.

### Error Categorization

This process involves grouping errors into specific types. Analysts investigate whether mistakes are concentrated in particular data subsets, correlate with certain input features, or are systematic versus random. This categorization helps prioritize the most impactful improvements.

### Failure Case Examination

This qualitative method involves the manual inspection of individual incorrect predictions, especially those where the model was highly confident. This can uncover challenging edge cases, errors in the data labels, or fundamental limitations of the model that are not apparent from summary statistics.

### Comparison to Baselines and Benchmarks

This final step puts the model's performance into context. Comparing results to simple baselines confirms that the model adds value, while comparison to state-of-the-art benchmarks or commercial systems indicates its competitiveness. Analyzing performance gaps helps identify areas requiring further development.

## 5.3 Hyperparameter Tuning

Hyperparameters are the configuration settings for a machine learning model that are not learned from the data but must be set beforehand. Finding the right combination of these settings is a crucial step to maximize model performance. This search for the optimal configuration is known as hyperparameter tuning or optimization.

## Types of Hyperparameters

Hyperparameters can be grouped into distinct categories based on which aspect of the learning process they govern. Key categories include settings that control the model's architecture, the speed and strategy of its learning algorithm, and the rules used to prevent overfitting.

### Model Architecture Hyperparameters

These settings define the model's structure and complexity. Examples include the depth of a decision tree, the number of layers in a neural network, or the number of neighbors considered in a similarity-based model. They control how complex a pattern the model can learn.

### Training Process Hyperparameters

These parameters control the learning process itself. Key examples are the *learning rate*, which sets the step size for updates; the batch size, which determines how many examples are processed at once; and the number of times the model goes through the full training dataset. They influence how effectively and efficiently the model learns.

### Regularization Hyperparameters

These settings are designed to prevent overfitting. They include the strength of penalty terms (L1/L2 regularization), the *dropout* rate in neural networks, and the intensity of data augmentation. They control the trade-off between learning the training data well and generalizing to new data.

## Tuning Strategies

Various strategies exist to search for the optimal hyperparameters, each with a different balance of computational cost and thoroughness. The goal of this automated process, known as hyperparameter tuning, is to find the configuration that results in the best model performance on a validation set.

### Manual Tuning

This approach relies on the practitioner's intuition and experience to adjust hyperparameters iteratively based on performance. While it can leverage deep domain knowledge, it does not scale well, and its success is highly dependent on the individual's expertise.

### Grid Search

This method performs an exhaustive search, training and evaluating a model for every possible combination of hyperparameters within a pre-defined set of values. It is thorough but becomes computationally prohibitive as the number of hyperparameters increases.

### Random Search

This strategy randomly samples hyperparameter combinations from defined ranges. It often finds good configurations more efficiently than grid search, especially when only a few hyperparameters significantly impact performance, as it better explores the entire search space.

### Bayesian Optimization

This is a more advanced strategy that builds a probabilistic model of the relationship between hyperparameters and performance. It uses this model to intelligently select the most promising hyperparameters to evaluate next, efficiently balancing the exploration of new areas with the exploitation of known good ones.

### Automated Hyperparameter Optimization

This approach uses specialized software tools to manage the entire search process. These tools automatically test different hyperparameter combinations, intelligently stop trials that are performing poorly, and run multiple tests in parallel. This provides full automation of the tuning workflow while still allowing for human guidance and oversight.

## Validation Methods

A robust validation methodology is essential during tuning to ensure that the selected hyperparameters generalize to unseen data. Without it, there is a high risk of overfitting the hyperparameters to the specific validation set, which will lead to poor performance in production.

### Nested Cross-Validation

This technique uses two layers of data splitting. An inner loop is used to tune the hyperparameters, while an outer loop provides an unbiased estimate of the model's performance on entirely held-out data. This prevents the hyperparameters from being overfit to a single validation set.

### Separate Validation and Test Sets

A fundamental practice is to maintain three distinct datasets: one for training, one for validation (to guide hyperparameter tuning), and one for testing (to provide a final, unbiased performance estimate after tuning is complete).

### Budget Constraints

In practice, hyperparameter tuning is constrained by limited computational resources and time. The goal is to find a set of hyperparameters that works well enough without wasting time or resources.

# 5.4 Ensuring Reliability and Safety Before Deployment

Reliable machine learning deployment depends not only on accuracy but also on robustness, transparency, and safety under real-world conditions. Before a model can be trusted in production, it must undergo rigorous validation and stress testing to ensure dependable performance across diverse scenarios. The following sections outline key evaluation methods used to confirm readiness for deployment.

## Validation and Testing

Validation and testing establish confidence in model readiness for deployment through comprehensive evaluation under realistic conditions. Functional testing verifies that a model and its supporting system operate as intended.

### Unit Tests

These tests examine individual components in isolation to ensure they work correctly. For example, they check if data is being cleaned properly, if the model's structure is built as designed, or if the error-calculation function returns the right value. This helps catch basic coding mistakes early.

### Integration Tests

These tests verify that all the different components of the system work together correctly. They ensure that the data pipeline correctly feeds the model, that the model's predictions are formatted properly for the next system, and that the entire workflow can handle different types of inputs without breaking.

### Regression Tests

The purpose of regression tests is to ensure that new improvements or changes to the model do not accidentally break existing functionality. After any modification, these tests are run to confirm that everything that worked previously still works as expected, thereby enabling confidence and continuous development.

## Performance Testing

Performance testing validates that the model meets specific requirements for accuracy, speed, and resource usage. This process is critical before deployment to ensure the system can handle real-world demand and deliver timely predictions without excessive computational cost.

### Accuracy Testing

This is the final check of the model's predictive power on a completely unseen dataset that represents real-world conditions. It confirms that the model has learned general patterns and can perform well on new data, including on important subgroups and challenging edge cases.

### Latency Testing

This test measures how long it takes for the model to make a single prediction. For applications that require instant results, latency must be extremely low, while for batch processing, longer times may be acceptable. This testing helps identify and fix speed bottlenecks.

### Throughput Testing

While latency measures the speed of one prediction, throughput measures how many predictions the system can handle at once. This is critical for services with many simultaneous users, as it tests the system's capacity under a heavy load.

### Resource Testing

This evaluates the computational cost of running the model, including its memory, processing power, and storage needs. Understanding these requirements is essential for deploying models on devices with limited capacity, like phones, or for managing cloud computing costs.

## Robustness Testing

Robustness testing evaluates how well a model performs when it encounters slightly altered or unexpected inputs. For example, small, almost invisible changes to an image or data point can sometimes cause a model to make incorrect predictions, a vulnerability known as adversarial sensitivity. Robustness testing checks whether the model can resist such subtle changes and continue to produce reliable results, an important step for ensuring trustworthiness in real-world applications. Adversarial robustness evaluation specifically tests how well a model resists small, intentional input changes using methods like FGSM or PGD that are designed to fool it.

### Noise Robustness

This tests the model with messy, incomplete, or slightly corrupted data, simulating the imperfections of the real world. A robust model should see a graceful decline in performance rather than a complete failure when faced with such inputs.

### Adversarial Robustness

This assesses the model's vulnerability to specially crafted inputs designed to trick it. These attacks work by making subtle, often imperceptible, alterations to legitimate data to cause incorrect predictions. While it is difficult to make a model completely immune to these attacks, this testing helps identify its biggest weaknesses.

### Distribution Shift Testing

This evaluates what happens when the data the model sees in production is different from the data it was trained on. This shift can occur over time, across geographic regions, or between different user groups, and is a common cause of performance decay.

### Edge Case Testing

This involves checking the model's behavior on rare, unusual, or extreme examples that were likely underrepresented in the training data. Identifying these failure points is the first step to addressing them through better data collection or safety mechanisms.

## Safety and Reliability Testing

This form of testing focuses on preventing harm from model errors or misuse. The process involves systematically probing for edge cases, stress-testing the model under extreme conditions, and ensuring that any failures it makes are predictable, understandable, and do not lead to harmful outcomes.

### Failure Mode Analysis

This is a proactive process of imagining how the model could fail and what the consequences of those failures might be. By systematically identifying potential risks, teams can develop strategies to mitigate them before deployment.

### Graceful Degradation

This principle ensures that when a model is uncertain or encounters a situation it cannot handle, it fails in a safe way. This could mean refusing to make a prediction, asking for human help, or defaulting to a safe, pre-defined action instead of making a risky guess.

### Redundancy and Validation

This involves adding extra layers of safety to catch errors. For high-stakes decisions, this could mean using two different models to cross-check each other, implementing rule-based systems to block obviously wrong outputs, or requiring human review. These steps increase system complexity and cost but are essential for safety.

### Human-in-the-Loop Systems

In high-risk AI applications—such as healthcare diagnostics, autonomous driving, or financial decision-making—complete automation can introduce unacceptable levels of risk. *Human-in-the-loop (HITL)* systems address this by ensuring that qualified humans remain actively involved in reviewing, approving, or intervening in critical AI outputs. This approach combines the efficiency of machine intelligence with human judgment, aligning with modern AI safety and governance standards that emphasize accountability, transparency, and responsible oversight.

# 5.5 Deploying Machine Learning Models to Production

Deployment transitions models from development to production, enabling real-world usage while managing operational complexities. This involves integrating the model into existing applications or services, establishing robust monitoring, and creating pipelines for reliable, scalable inference. Successful deployment ensures that the model not only performs well in controlled testing but also remains stable and effective under real-world conditions.

## Deployment Strategies

Different deployment strategies balance risk, complexity, and rollout speed. The choice of strategy is crucial for minimizing disruption, gathering performance data safely, and ensuring a smooth user experience during updates.

### Shadow Deployment

This strategy involves operating a new model in parallel with the existing production system without directing any live traffic to it. The new model receives the same input data and generates predictions, which are logged and compared against the legacy system's output for evaluation. This "shadow mode" validates performance and stability under real-world load without operational risk, though it cannot test the full impact on downstream systems or end-user experience.

## Canary Deployment

This strategy involves gradually rolling out a new model to a small, specific subset of users before a full-scale deployment. Releasing it to just 1–5% of traffic lets teams monitor performance in the live environment without affecting most users. If the canary release is successful, the new model is then progressively expanded to larger user populations.

## Blue-Green Deployment

This strategy maintains two identical, separate production environments: "blue" (running the current stable version) and "green" (hosting the new candidate version). After the green environment is fully validated, all user traffic is seamlessly switched from blue to green. This approach enables instant rollback by simply redirecting traffic back to the blue environment if any issues are detected, effectively minimizing downtime and risk at the cost of maintaining duplicate infrastructure.

## A/B Testing

This method compares two versions of a model or feature by showing each one to different groups of users and measuring which performs better. Statistical analysis helps determine whether the difference in performance is real or just due to chance. By offering a clear, evidence-based comparison in real-world conditions, A/B testing helps teams choose the version that best meets business goals. It remains the most trusted method for measuring the true impact of a change before a full rollout.

## Continuous Integration, Delivery, and Monitoring

Modern machine learning systems use continuous integration and delivery (CI/CD) pipelines to automate testing, updating, and deployment. These pipelines ensure that models move smoothly from development to production while maintaining consistency and reliability. Just as important is continuous monitoring, which tracks model performance to detect issues such as data drift, when real-world data begins to differ from the training data. When performance declines, automated alerts or retraining triggers can update the model, helping sustain accuracy, fairness, and long-term stability in production environments.

# MLOps: Managing the Machine Learning Lifecycle

### Foundations and Automation

The deployment of a model is not the end of its journey—it marks the beginning of an ongoing operational cycle. MLOps (Machine Learning Operations) extends DevOps principles to machine learning systems by integrating development, deployment, and monitoring into a unified, automated workflow. Through continuous integration and delivery (CI/CD), models can be updated, tested, and redeployed efficiently.

### Monitoring and Continuous Improvement

A key focus of MLOps is model performance monitoring, which tracks metrics over time to detect data drift or concept drift—indicators that real-world data is changing. When drift or performance degradation is detected, automated retraining pipelines can refresh the model using new data, ensuring that predictions remain accurate and fair.

Beyond automation, MLOps emphasizes collaboration, version control, and governance, providing traceability from raw data to deployed models. This discipline enables organizations to scale machine learning responsibly, maintain reliability, and ensure continuous alignment between evolving data, business goals, and ethical standards.

# Infrastructure Requirements

For a machine learning model to work reliably in the real world, it needs a robust technical foundation, much like a website needs a reliable web server. This infrastructure is responsible for serving predictions, managing computational resources, and handling user traffic. Without this supporting architecture, even the most accurate model is merely a theoretical artifact, unable to deliver value in a production environment.

### Serving Infrastructure

This is the system that receives requests and delivers predictions. It must be built to handle many users at once, respond quickly, and stay online 24/7. It automatically adjusts its capacity during busy periods and uses smart techniques to avoid unnecessary work, ensuring a smooth experience for everyone.

### Model Storage and Versioning

Just like version history is used in a document, teams need a system to store, track, and manage different versions of their models. This model registry keeps a record of which data was used to train each model and how well it performed. This allows teams to easily roll back to a previous version if a new model causes problems.

### Dependency Management

A model relies on specific software to run. This part of the infrastructure ensures that the model has access to the exact same software environment it was built with, no matter where it's deployed. This prevents unexpected crashes and ensures the model behaves consistently.

### Monitoring Infrastructure

Once a model is live, its health and performance need to be monitored. This system acts like a dashboard and alarm system, tracking everything from speed and accuracy to system errors. It alerts the team immediately if something goes wrong, allowing for a quick response.

## Integration

A model's value is realized only when its predictions are seamlessly woven into other software and user experiences. This is achieved by exposing the model's functionality through an API, which allows various applications to request predictions on demand. This integration transforms a standalone analytical asset into an active component that can automate decisions, personalize content, or generate insights within a live product.

### API Design

An API is a standard way for other software programs to ask the model for a prediction. Good design makes this communication simple, fast, and reliable, allowing other applications to easily use the model's intelligence.

### Data Pipelines

Models need a constant supply of fresh data to make predictions. These pipelines are automated pathways that collect, prepare, and deliver this data from its source, such as user activity or a database, to the model, ensuring it always has the right information to work with.

### Application Integration

This final step operationalizes the model's predictions to drive decisions and actions. The output might power a product recommendation on a website, flag a fraudulent transaction in real-time, or assist a doctor with a diagnostic insight. Successful integration is achieved when this insight is delivered seamlessly, at the right moment and in a format that is immediately useful and actionable for the end-user.

Machine learning practice is evolving toward MLOps – the integration of machine learning with DevOps principles. This approach ensures reproducibility, version control, automated testing, and continuous deployment of models, enabling scalable and reliable AI systems.

Chapter 6

# Monitoring, Feedback, and Continuous Improvement

## 6.1 Designing for Real Users

A machine learning system's journey does not end at deployment; it enters a critical new phase where its real-world performance and impact must be actively managed. This chapter explores the essential disciplines of monitoring for model degradation, establishing feedback loops to capture new data, and designing systems with the end-user in mind, all to enable continuous adaptation and long-term success.

Machine learning systems achieve their true value only when they integrate seamlessly into human workflows. Designing for real users means understanding their goals, constraints, and mental models. This requires close collaboration among data scientists, UX designers, and domain experts to ensure that model outputs are not just accurate, but are presented as meaningful and actionable insights.

The ultimate goal is to create intuitive interfaces and clear outputs that users can trust and act upon with confidence. By anchoring technical excellence in user-centered design, machine learning systems become tools for empowerment rather than sources of complexity.

## 6.2 User Interface and User Experience

### Designing Interfaces

The success of a machine learning system in real-world applications depends as much on its interface as on its algorithmic sophistication. No matter how advanced the model, users judge the system through what they can see, understand, and control. A well-designed interface transforms raw model output into actionable insight; a poorly designed one can undermine trust, obscure limitations, and lead to misuse.

User experience (UX) in machine learning goes beyond looks or basic usability. It encompasses how the system communicates its reasoning, conveys uncertainty, supports user judgment, and adapts to feedback. The interface acts as the bridge between statistical intelligence and human decision-making, translating abstract probabilities into meaningful, context-aware guidance.

A thoughtful UX design must therefore achieve four intertwined goals:

- **Clarity:** The system's purpose, outputs, and limitations should be evident. Users should immediately understand what AI is doing, what each output means, and how they are expected to respond. Avoiding technical jargon and showing results in clear formats, like graphs, confidence bars, or ranked lists, helps users understand them correctly.
- **Transparency:** Machine learning often operates as a "black box." Interfaces can counter this opacity by revealing how inputs influence predictions or by highlighting relevant factors behind a decision. Even limited transparency, such as feature importance indicators or plain-language rationales, builds comprehension and trust.
- **Trust Calibration:** Good UX guides users to rely on the system appropriately, avoiding both blind trust and unnecessary doubt. This requires signaling confidence levels, clarifying when predictions are uncertain, and giving users the ability to question or override automated results.
- **Feedback and Learning:** Interfaces should make it easy for users to provide feedback, correct errors, or express preferences. These mechanisms not only improve user engagement but also supply valuable data for model retraining and refinement, closing the loop between human and machine learning.

## Ethical Design Considerations

In high-stakes domains such as healthcare, finance, or criminal justice, UX design becomes an ethical responsibility. Poorly explained predictions can mislead users, and overly confident interfaces can hide uncertainty. In consumer applications, thoughtful design prevents frustration, mitigates bias, and promotes fair outcomes across user groups.

Ultimately, the user interface is where technical intelligence becomes human-centered intelligence. It transforms the model's internal logic into experiences that are understandable, trustworthy, and actionable, determining whether a machine learning system is not only accurate but also adopted and valued in practice.

## 6.3 Human-AI Interaction

Machine learning systems ultimately succeed or fail based on how humans perceive, understand, and interact with them. The design of interfaces and interaction patterns determines whether AI outputs are clear, trustworthy, and useful in practice. This section explores two complementary aspects of this relationship: how results are presented (Interface Design) and how users engage with the system (User Interaction Patterns).

### Interface Design

The design of the interface determines how users receive and interpret the system's outputs. Effective design translates complex probabilities and classifications into clear, contextualized information, such as confidence scores, visual highlights, or plain-language recommendations. The goal is to build appropriate trust, prevent misinterpretation, and provide users with a clear path to action, ensuring the model's intelligence is accessible and useful rather than confusing or overwhelming.

#### Prediction Presentation

Model outputs must be communicated with clarity. This may involve presenting a simple classification label accompanied by a confidence score, a numerical prediction with an uncertainty range, or an ordered list for ranking tasks. The use of visual aids, such as charts or highlighted text, can make complex information more accessible. The presentation style should be tailored to the user's level of expertise and their specific decision-making context.

### Explanation and Transparency

To build trust and enable informed action, systems should provide insight into their reasoning. This can be achieved by showing which input factors most influenced a decision, by presenting similar historical cases, or by offering a natural language summary of the logic. Appropriate transparency helps users understand not just what the system decided, but why.

### Confidence Communication

It is essential that the interface conveys the reliability of its own predictions. This can be accomplished through numerical probability scores, or through visual cues like color-coding and icons that signal confidence levels. This helps users see when a prediction is reliable and when it needs closer scrutiny.

### Error Handling

Interfaces must be designed to manage situations where the model cannot provide a confident or reliable output. In such cases, the system should gracefully communicate its limitations, suggest alternative actions, or defer to human judgment, rather than presenting a potentially incorrect prediction as fact.

## User Interaction Patterns

Different applications call for different patterns of interaction between the user and the machine learning system. For instance, a music streaming service might use a passive pattern, where recommendations are generated automatically based on user behavior, while a data analysis tool might use a reactive pattern, where the user submits a query and the model generates a specific insight on demand. The choice of pattern is dictated by the application's goal—whether to proactively guide the user, augment their decision-making, or respond directly to their requests.

### Automated Decisions

In this pattern, the system applies predictions without direct human involvement. Examples include spam filtering and content recommendation engines. This approach is highly scalable but requires a high degree of inherent reliability and safety assurance.

### Decision Support

In this interaction pattern, the system provides predictions to a human expert, who retains authority over the final decision. This pattern is valuable in high-stakes domains like medicine or law, as it combines the analytical power of AI with human expertise, judgment, and accountability.

### Interactive Refinement

This pattern establishes a feedback loop in which user input is utilized to directly refine and enhance the model's subsequent predictions. Through actions such as correcting errors, adjusting parameters, or indicating preferences, users iteratively steer the model's behavior. This process facilitates a collaborative and adaptive interaction, progressively aligning the system's outputs with the user's specific intent and context.

### Exploration and Discovery

In this mode, machine learning serves as a tool for insight, helping users uncover patterns, visualize complex data, or identify anomalies. These interfaces support insight development and hypothesis generation, empowering users to discover new knowledge within their data. For instance, a financial analyst might use such a tool to visually explore clusters of transactions and identify previously unknown patterns of fraud.

## 6.4 Building Trust

User trust is the critical determinant of whether a machine learning system will be adopted and relied upon in practice. This trust is not automatically granted; it must be intentionally built through transparent design, reliable performance, and clear communication. Establishing it requires demonstrating competence through accuracy, fostering understanding through explainability, and ensuring ethical alignment through fairness and accountability.

Trust in AI systems emerges from two interconnected dimensions: system integrity (how reliably, transparently, and fairly the system operates) and user empowerment (how much control and confidence the user feels when engaging with it).

## System Integrity

### Reliability and Consistency

Trust is built when a system demonstrates dependable performance over time. Users must be able to count on consistent and accurate outputs. Systems that make frequent or unpredictable mistakes quickly lose user trust, underscoring the necessity of rigorous testing and quality assurance prior to deployment.

### Transparency

Trust is fostered when systems are open about their processes. This involves explaining the rationale behind decisions, openly acknowledging system limitations, and clearly communicating uncertainty. Black-box systems that give no insight into how they work naturally create skepticism and weaken trust.

### Fairness

A fundamental component of trust is the assurance that a system treats all user groups equitably. Systems that exhibit bias or produce discriminatory outcomes against any demographic group will justifiably destroy trust. Therefore, proactive testing for bias and a demonstrated commitment to mitigation are essential. This commitment must be continuous, extending from the initial design phase through to post-deployment monitoring, to address biases that may emerge over time.

## User Empowerment

The second dimension of trust focuses on the user's sense of control and comfort when interacting with the system. Even the most accurate model will fail to gain acceptance if users feel disempowered or alienated by it.

### Control and Agency

Trust is strengthened when users feel a sense of control. Providing options to override a system's recommendation, customize its behavior, or opt-out of its use respects user autonomy and signals that the system is a tool to augment, not replace, human judgment. Forcing reliance on automated decisions, especially without explanation, often leads to user frustration and resistance. This collaborative approach transforms the user's relationship with the system from one of passive submission to active partnership, which is the bedrock of sustainable trust.

### Gradual Onboarding

Gradually introducing users to an AI system helps build confidence over time. Beginning with simple, highly explainable features allows users to experience the system's value and build a foundation of trust before more complex capabilities are introduced. This structured onboarding process reduces the cognitive load and perceived risk for the user. Ultimately, this phased approach fosters a sense of mastery and ensures that the user's initial interactions are successful and positive, paving the way for long-term adoption.

# 6.5 Feedback and Continuous Improvement

Machine learning systems are not static artifacts; their performance can degrade as the world changes through data drift, evolving user behavior, or shifting real-world conditions. A structured, operational process for gathering feedback—from both explicit user input and implicit performance metrics—is therefore necessary to maintain their relevance and accuracy over time. This cyclical process of monitoring, retraining, and redeployment transforms a static model into a living, adaptive system that can sustain its value long after its initial launch.

## Feedback Collection

Actively gathering information on how a model is performing in the real world is the first step toward improvement. This involves implementing robust monitoring to track key performance metrics, such as prediction accuracy and data drift, as well as creating channels for users to report errors or provide direct feedback on predictions. The goal of this data collection is to move from assumptions about how the system *should* work to evidence-based insights about how it *actually* performs, thereby identifying specific areas for refinement and adaptation.

### Explicit Feedback

This is feedback users give directly, such as a thumbs-up/down, a star rating, or a corrected prediction. It is very valuable because it is clear and explicit, though users may provide it only occasionally.

### Implicit Feedback

This type of feedback is inferred from user behavior rather than directly requested. Examples include what a user clicks on, how long they engage with content, or whether they successfully complete a task. While this data is abundant and scales easily, it requires careful interpretation to accurately understand user intent.

### Active Learning

This is a strategic approach to data collection where the system itself identifies which new data points would be most informative for its own improvement and proactively requests labels for them, making the learning process more efficient.

### Error Reporting

Providing users with simple, accessible channels—like a "report this" button—to flag incorrect outputs is crucial. These structured reports provide direct evidence of system failures and are invaluable for diagnosis, as they capture edge cases and real-world complexities that may be absent from the original training data. By systematically collecting and analyzing this feedback, development teams can prioritize the most impactful fixes and create targeted datasets for model retraining. This process not only improves the system's accuracy but also actively demonstrates to users that their input is valued, thereby reinforcing trust and fostering a collaborative relationship.

## Performance Monitoring

Continuously tracking key metrics is essential for catching performance issues early. This involves monitoring for both technical service-level indicators, like latency and throughput, and core model-health metrics, such as prediction accuracy and data drift. By establishing automated alerts for significant deviations from baseline performance, teams can proactively identify and diagnose problems like concept drift or model staleness before they significantly impact the user experience or business outcomes.

### Accuracy Monitoring

This process tracks key performance metrics like precision and recall over time to detect model degradation. A consistent decline indicates model drift, where live data diverges from the original training data. Automated monitoring against a validated baseline enables proactive retraining before performance issues impact users or business operations.

### Prediction Distribution Monitoring

This technique monitors the nature of the model's outputs themselves. A significant shift in the pattern of its predictions can signal that the model is behaving differently than expected, often due to changes in the incoming data. This kind of monitoring is important for spotting problems even when the true labels are delayed or unavailable.

### Input Distribution Monitoring

This tracks the characteristics of the data the model receives. If this "input data" begins to differ significantly from the data the model was originally trained on, a phenomenon known as "drift," it serves as an early warning that model performance is at risk.

### Operational Metrics

Beyond predictive performance, it is vital to monitor the system's technical health, including its speed (latency), error rates, and resource consumption. Deterioration in these areas can indicate infrastructure problems that need resolution. Establishing service-level objectives (SLOs) for these metrics ensures that the system meets the reliability and responsiveness requirements for a positive user experience.

## Model Updates

To maintain performance in a changing environment, machine learning models require regular updates. The frequency of these updates, known as the retraining cadence, must be calibrated to the domain. Rapidly evolving situations may necessitate daily updates, while more stable contexts may only require quarterly updates. This schedule balances the need for accuracy with the computational cost and operational effort involved.

### Incremental Learning

Instead of full retraining, models can sometimes be updated with new data alone. This approach is more efficient but requires careful management to ensure the model does not forget previously learned information or accumulate new errors over time.

### Data Refresh Strategies

A key decision is which data to use for retraining. Strategies range from using only the most recent data to adapt to new trends, to using all historical data to preserve long-term knowledge. The choice impacts both the model's relevance and the computational resources required.

### Validation of Updates

Every new model version must be rigorously tested before full deployment. This usually involves testing the old and new models in parallel on a small portion of traffic. Crucially, a rollback plan must be in place to quickly revert to the previous version if the new one underperforms.

## Continuous Improvement Process

A systematic process is required to translate feedback and experience into tangible enhancements. This begins by consolidating inputs from monitoring, error reports, and user feedback into a prioritized backlog of improvement tasks. These tasks then fuel an iterative cycle of retraining models on new data, refining features, and deploying updated versions, thereby transforming insights into action and ensuring the system evolves alongside its environment.

### Issue Triage

Not all problems are equal. This process involves categorizing and prioritizing identified issues based on their impact and frequency, ensuring that resources are allocated to the most critical improvements first. Impact is typically measured by the severity of the consequence, such as a model error causing a significant financial loss or a critical user-facing failure. This disciplined approach prevents teams from being overwhelmed and ensures that the most harmful flaws are addressed promptly.

### Root Cause Analysis

When a problem occurs, the goal is to find its true source–whether in the data, the model's design, or the deployment system–rather than only fixing the visible symptom. This helps prevent the issue from returning. Two simple tools can support this process. The '5 Whys' involves asking 'why?' several times to dig deeper into the cause, and fishbone diagrams show possible causes in a clear visual map. These methods help identify the root cause so the system becomes more reliable over time.

### Experimentation Framework

A structured environment for testing hypotheses is essential. This framework allows teams to safely trial new ideas, such as through A/B tests, and to learn rapidly from both successes and failures. By defining clear success metrics and statistical significance thresholds upfront, teams can make objective, data-driven decisions about which changes to implement permanently. This systematic approach de-risks innovation and ensures that continuous improvement is guided by evidence rather than intuition.

### Documentation of Learnings

This practice captures critical insights from production experience, asking what worked, what failed, and why. Documented learnings guide future decisions, preserve institutional knowledge, and prevent repeated mistakes. This creates a valuable, searchable knowledge base that accelerates the onboarding of new team members and improves the entire development lifecycle.

## Organizational Discipline

Continuous improvement should be viewed not just as a maintenance function but as an organizational discipline. Formal review cycles, post-mortem analyses, and scheduled performance audits create feedback loops that sustain long-term quality. When combined with transparent documentation and shared learnings, this process transforms the machine learning lifecycle into a living system – one that evolves in step with changing data, business goals, and ethical expectations.

# 6.6 Monitoring and Maintenance

Once deployed, machine learning systems require ongoing oversight to ensure they remain reliable and valuable over time. This involves establishing a comprehensive monitoring system to track not just the model's predictive performance, but also the health of the data pipelines and infrastructure that support it. Without this continuous vigilance, models are susceptible to silent failures, such as performance degradation from data drift or breakdowns in upstream data sources, which can erode user trust and business value before the issues become apparent.

## System Health Monitoring

A comprehensive monitoring system tracks the health of the entire application, from the underlying infrastructure to the data pipelines and the model serving endpoints. This includes monitoring standard operational metrics like CPU and memory usage, API latency, and error rates to ensure the system stays responsive and available. By establishing alerts based on these metrics, teams can be notified proactively of system failures or performance degradation, enabling a rapid response to minimize downtime and user impact.

### Application Metrics

These track the core functionality of the system from a user's perspective. Key indicators include request volume, error rates, response times, and success rates, which collectively reveal whether the system is serving its fundamental purpose effectively.

### Infrastructure Metrics

These measure the health and utilization of the underlying hardware and platform. Monitoring central processing unit (CPU) and memory usage, disk space, and network activity is essential, as constraints in these areas can directly impact the stability and speed of the entire application.

### ML-specific Metrics

These track the unique behavior of the machine learning model itself. This includes monitoring the distribution of its prediction confidence scores and the statistical properties of its outputs. Significant changes in these patterns can serve as an early warning of model degradation or data quality issues.

### Business Metrics

These connect technical performance to organizational objectives. By tracking key performance indicators (KPIs) such as user engagement, conversion rates, or revenue impact, teams can verify that the technical system is delivering tangible business value.

## Alerting and Incident Response

Effective alerting notifies the right people of problems, while a structured incident response process ensures they are resolved quickly. These alerts must be clear and useful, filtering out noise so developers don't get overwhelmed and miss real problems. Furthermore, the incident response process should include clear protocols for communication, escalation, and post-incident analysis to continuously improve system resilience.

### Alert Configuration

This involves defining clear rules for when notifications are sent. Alerts should be based on meaningful thresholds, be actionable by specifying the required intervention, and be routed to the appropriate team to avoid unnecessary noise and ensure a prompt response.

### Incident Response Procedures

These are predefined plans that guide the team's reaction when an alert is triggered. These procedures, often documented in runbooks–step-by-step guides for diagnosing and fixing issues–help reduce downtime and user impact.

### Post-Incident Reviews

After an incident is resolved, a blameless analysis is conducted to learn from the event. The focus is on understanding the root cause and identifying systemic improvements to prevent the problem from recurring, rather than on assigning individual faults.

## Maintenance Activities

Regular maintenance is required to preserve system health and performance in the long term. This includes proactive tasks such as applying security patches, updating dependencies, and rotating credentials to mitigate vulnerabilities. A well-documented maintenance schedule, often executed during predefined change windows, helps to minimize operational disruption while ensuring consistency.

### Dependency Updates

This involves keeping all software libraries, frameworks, and system components current. Regular updates are crucial for applying security patches, accessing bug fixes, and incorporating performance improvements, and are typically scheduled to minimize user disruption.

### Performance Optimization

This ongoing activity aims to maintain system efficiency. It can include refining database queries, improving code structure, and adjusting resource allocation to prevent the gradual performance degradation that can occur over time.

### Technical Debt Management

This is the process of addressing shortcuts and compromises taken during development. By periodically streamlining the underlying software, improving architecture, and updating documentation, teams prevent the system from becoming brittle, complex, and difficult to maintain. This proactive care reduces the accumulation of small issues that can later cause major outages or slow down the development of new features. Ultimately, it is a crucial investment in the long-term health and agility of the system, ensuring it can continue to evolve to meet business needs.

### Capacity Planning

This proactive process involves forecasting future demand based on growth trends. Its goal is to ensure that sufficient computational resources are available to meet user needs, enabling cost-effective scaling and preventing service interruptions due to overload.

## 6.7 Scalability

Scalable machine learning systems are designed to handle growth efficiently, whether in data volume, number of users, or computational demands. This involves designing components that can be easily duplicated (horizontal scaling) or upgraded (vertical scaling) to manage increased load without a complete redesign. For instance, a scalable system can seamlessly add more servers to process a surge in data or distribute the workload across multiple machines to maintain fast response times for users. This elasticity ensures that performance remains consistent and cost-effective as requirements evolve.

### Computational Scalability

This ensures the system can process large amounts of data and serve many users without slowing down. This is typically achieved by designing systems that can distribute workloads across multiple processors or machines, such as through a cloud-based infrastructure. This elastic approach allows resources to be scaled up during peak demand and scaled down during quieter periods, optimizing both performance and cost.

#### Training Scalability

This involves using multiple processors or machines to train a model on a massive dataset. By splitting the work, the system can learn from enormous amounts of information in a reasonable time, making large-scale model development feasible.

#### Inference Scalability

This focuses on delivering predictions quickly to a large number of users. Techniques include adding more servers to handle traffic, storing common results to avoid recalculating them, and simplifying the model to make it faster, all while maintaining low response times.

### Resource Optimization

The goal is to maximize the efficiency of available hardware resources, ensuring that powerful graphics processors and other computing units are fully utilized. This approach helps control operational costs by extracting the greatest possible value from the infrastructure being used.

## Data Scalability

This addresses the challenges of managing and processing ever-growing datasets. It requires robust data storage solutions and pipelines that can handle huge amounts of data quickly without slowing the system down. Effective data scalability ensures that as the volume of data expands, the time to train models and generate insights remains practical, preventing data growth from crippling the entire machine learning operation.

### Storage Scalability

This involves using distributed systems to store massive datasets reliably. A common strategy is "tiered storage," which keeps frequently accessed ("hot") data on fast, expensive storage and less used ("cold") data on cheaper, slower storage. This automated approach optimizes both performance and cost, ensuring that resources are allocated efficiently based on the actual value and usage patterns of the data.

### Processing Scalability

To analyze large datasets, the work is broken down and processed in parallel across many computers. For continuous data, streaming systems can process information as it arrives, enabling real-time insights. This parallel approach drastically reduces the time required for complex computations, transforming tasks that would take days on a single machine into ones that can be completed in hours or minutes.

### Feature Engineering at Scale

This involves building systems that can reliably and efficiently compute the input features for models from huge datasets, often by pre-calculating and storing them for quick access. This ensures that models are served with consistent, up-to-date features without becoming a bottleneck during training or real-time prediction.

## Operational Scalability

This ensures that the processes for developing and managing models remain effective as the number of users and projects grows. This is achieved by implementing standardized workflows, automated testing, and centralized model registries to maintain order and reproducibility. Without this structure, teams can get stuck dealing with coordination issues and inconsistent practices, slowing innovation and deployment.

### Team Scalability

This is achieved by building systems with clear, modular components, allowing multiple teams to work independently without interfering with each other's work. This modularity establishes well-defined interfaces between different parts of the system, such as data sources and feature stores, which minimizes dependencies and coordination overhead. By empowering teams to develop, deploy, and own their services, the organization can scale its engineering efforts efficiently without creating a bottleneck.

### Experiment Scalability

This involves using platforms that allow data scientists to run many tests and A/B trials simultaneously in a controlled way, accelerating the pace of innovation without causing chaos. These platforms provide standardized environments and resource management to ensure that experiments are reproducible and their results are comparable. This structured approach to experimentation is crucial for systematically exploring new ideas and reliably identifying the most promising models for production.

### Model Management at Scale

As the number of models grows, a systematic approach is needed. This involves using a central "model registry" to track versions and automated pipelines for deployment, preventing confusion and ensuring reliability. The registry acts as a single source of truth for model lineage, performance metrics, and stage (e.g., staging, production), while automated CI/CD pipelines for machine learning ensure consistent, reliable, and rapid model deployments. This robust governance is essential for auditing, rollback, and maintaining user trust as the model inventory expands.

## 6.8 Data Security and Privacy

Machine learning systems that handle sensitive information must be designed to protect it and comply with regulations. This involves implementing encryption, strict access controls, and auditing to prevent unauthorized access. For stronger privacy, techniques like federated learning can help. Instead of sending all the data to one place, federated learning trains the model where the data already lives and only shares the results of that training, not the raw information. A proactive stance on security is essential for maintaining user trust and operational integrity.

### Data Protection

These are the technical measures used to safeguard data. This includes encrypting data both when it is stored and when it is being transmitted between systems. Additionally, strict access controls and authentication mechanisms ensure that only authorized users and services can retrieve or modify sensitive information. Regular security audits and monitoring are also essential to detect and respond to potential vulnerabilities or breaches in a timely manner.

### Access Control

This ensures that only authorized personnel can view or use data, following the principle of granting the minimum level of access necessary to perform a job. This is often managed through role-based permissions, which automatically assign access rights based on a user's defined responsibilities within the organization.

### Encryption

This protects data by converting it into a secure code, both when it is stored ("at rest") and when it is being transmitted ("in transit"). Without the correct cryptographic keys, the encrypted data remains an unreadable and useless string of characters, even if it is intercepted or stolen.

### Data Anonymization

This involves techniques to remove or obscure personal identifiers from a dataset, allowing the data to be used for analysis while protecting individual privacy. However, it requires careful implementation, as sophisticated techniques can sometimes re-identify individuals if not done thoroughly.

### Secure Computation

These are advanced techniques that allow a model to be trained on data without ever centralizing it or decrypting it, enabling collaboration on sensitive information. Methods like federated learning, for example, send the model to the data instead of bringing the data to the model, thus keeping the raw information securely decentralized.

## Privacy Compliance

This involves adhering to laws and regulations that govern the handling of personal data. Organizations must implement clear policies and procedures to honor user rights, such as the ability to access, correct, or delete their personal information. This often requires maintaining detailed records of data processing activities to demonstrate compliance during audits and building trust by showing a commitment to responsible data stewardship.

### Regulatory Frameworks

Laws like GDPR in Europe or CCPA in California set the rules for data collection and usage, granting individuals rights over their personal information. Non-compliance can result in significant financial penalties and reputational damage for an organization.

### Data Minimization

This is the practice of only collecting the data that is absolutely necessary for a specific purpose and deleting it when it is no longer needed, which reduces privacy risks. This principle is fundamental to building systems that are secure and respectful of user privacy by design.

### Consent Management

This involves systems that clearly ask users for permission to use their data, explain how it will be used, and provide a way for users to withdraw their consent. A robust system must reliably log and track user preferences to ensure that all data processing aligns with the granted permissions.

### User Rights Support

This refers to the technical and operational capability to fulfill user requests, such as providing a copy of their data, correcting it, or deleting it. Automating these processes is often essential for responding to requests accurately and within the legally mandated timeframes.

## Model Security

This focuses on protecting the machine learning models themselves from threats. A key concern is adversarial attacks, where someone intentionally alters an input, often in tiny, hard-to-notice ways, to trick a model into making the wrong prediction. To defend against this, models must be rigorously tested for vulnerabilities and monitored in production for unusual activity. Additionally, access to the models themselves must be controlled to prevent theft or unauthorized modification, safeguarding valuable intellectual property and ensuring system integrity.

### Model Extraction

This is the risk of someone stealing a proprietary model by repeatedly querying it. Defenses include limiting query rates and embedding hidden markers to prove ownership. This type of attack aims to replicate a model's functionality, effectively devaluing the original investment in its development.

### Adversarial Attacks

These are specialized inputs designed to trick a model into making a mistake. Defenses include training the model on similar "trick" examples to make it more robust. These attacks can be as subtle as adding a layer of digital "noise" to an image that is invisible to humans but causes the model to misclassify it entirely.

### Model Poisoning

This occurs when an attacker corrupts the training data to compromise the model's performance. Defenses involve carefully validating data and using robust training algorithms. A poisoned model may perform well during testing but fail or behave maliciously when deployed in the real world.

### Intellectual Property Protection

This involves using a combination of legal tools (like patents) and technical measures (like code obfuscation) to prevent the unauthorized use or duplication of a valuable model. This is crucial for maintaining a competitive advantage and ensuring that the significant resources spent on research and training are protected.

## 6.9 Ethical Considerations

Machine learning systems have a significant impact on people and society, creating an ethical responsibility for developers and organizations that goes beyond simply following the law. This involves proactively identifying and mitigating potential harms, such as the unfair exclusion of certain groups or the erosion of personal autonomy. Establishing transparent processes and clear accountability for an AI system's outcomes is essential for building public trust and ensuring these powerful tools are used responsibly.

## Fairness and Bias

The goal of fairness is to ensure that automated systems treat all individuals and groups equitably. This requires actively testing for and mitigating biases that can be inherited from historical training data or introduced through model design. Achieving this involves using specialized metrics and techniques to evaluate a model's performance across different demographics, ensuring it does not produce disproportionately adverse outcomes for any particular group. A continuous commitment to fairness is necessary to build trustworthy and just AI systems.

### Bias Sources

Bias can enter a system in several ways. It can be present in the historical data used for training, which may reflect past discrimination. It can occur if certain groups are underrepresented in the data, or if the way a characteristic is measured is itself flawed. Understanding these sources is the first step toward fixing them.

### Fairness Metrics

Fairness can be defined and measured in different ways, and these definitions can sometimes conflict. For example, one metric might require the same acceptance rate for all groups, while another might require the same accuracy in identifying qualified candidates. The choice of which metric to prioritize depends on the specific context and the values of the organization.

### Bias Mitigation

There are techniques to reduce unfair bias. These can involve cleaning the training data, building fairness constraints directly into the learning algorithm, or adjusting the model's predictions after they are made. While no method can eliminate bias completely, these approaches can significantly reduce it.

### Fairness Auditing

This is the ongoing process of testing a system for discriminatory outcomes. It involves breaking down performance metrics by different demographic groups to check for unequal impacts. Regular audits are necessary to catch biases that may emerge as the system operates in the real world. When disparities are identified, the underlying data or model must be adjusted to promote more equitable results.

## Transparency and Accountability

These principles ensure that someone is responsible for the system's behavior and that its workings can be understood. Transparency is often achieved through explainable AI (XAI) techniques, which provide human-interpretable reasons for a model's decisions. Clear accountability involves defining who is responsible for a model's development, deployment, and ongoing monitoring, ensuring that any negative impacts can be promptly addressed. This framework is crucial for auditing outcomes, maintaining regulatory compliance, and fostering user trust.

### Documentation

This involves creating clear records about a system's purpose, how it was built, its limitations, and its known flaws. This documentation, such as "model cards," helps users and regulators make informed decisions about whether and how to use the system.

### Explainability

This is the ability to understand why a model made a particular decision. Some simpler models are inherently interpretable. For complex models, special techniques can be used to highlight which factors most influenced a specific outcome, which is crucial for building trust and debugging errors.

### Auditability

This involves designing systems so their decisions and underlying processes can be reviewed and verified later. It requires maintaining detailed logs and version histories of both code and data, allowing independent review and investigation when issues arise.

### Accountability Mechanisms

This involves clearly assigning responsibility for the system's outcomes. It includes establishing governance committees for high-stakes applications and creating channels for individuals to appeal automated decisions that affect them, ensuring there is a human answerable for any harm that is caused.

## Social Impact

Machine learning systems can have profound and long-term effects on society that must be carefully considered. This includes evaluating potential consequences such as the displacement of workers through automation or the amplification of societal divides via biased algorithms. Proactively engaging with ethicists, policymakers, and the communities affected by these technologies is crucial for guiding their responsible development. Ultimately, the goal is to ensure that machine learning systems are designed and deployed in a way that benefits society as a whole and aligns with broader human values.

### Harm Prevention

This is a proactive effort to identify who could be negatively affected by a system and in what ways. Through systematic risk assessments and testing under diverse conditions, developers can introduce safeguards before deployment to minimize or prevent harm.

### Stakeholder Engagement

This involves actively including people from the communities affected by the technology in the design process. Their input can reveal unintended consequences and ethical concerns that developers might otherwise overlook, resulting in more equitable and widely accepted outcomes.

### Value Alignment

This pushes developers to ask a critical question: Whose values are being programmed into the system? Making these value judgments explicit, rather than allowing them to be hidden by technical choices, ensures the system's behavior aligns with ethical principles.

### Long-Term Impacts

This principle encourages looking beyond short-term gains to consider how a system may influence society over time – including its effects on jobs, social relationships, and power dynamics. Taking a long-term perspective helps prevent solutions that solve today's problems but create larger ones tomorrow. Such foresight is essential to guide technological progress toward outcomes that are sustainable and equitable for future generations.

## 6.10 Documentation and Reporting

Clear and comprehensive documentation is essential for others to understand, use, maintain, and trust a machine learning system. This includes detailing the model's intended use, its limitations, the data it was trained on, and its performance characteristics. Effective reporting also involves creating clear summaries of model behavior and business impact for stakeholders who may not have a technical background. Maintaining this practice ensures knowledge is preserved over time and facilitates collaboration across different teams.

### Technical Documentation

This type of documentation explains how the system is built and how it works. It typically includes architecture diagrams, code comments, API specifications, and descriptions of the data pipelines. This level of detail is crucial for engineers to debug issues, onboard new team members, and make future modifications safely and efficiently. Without it, systems become "black boxes" that are difficult to maintain or improve.

### Architecture Documentation

This provides a high-level "blueprint" of the system. It uses diagrams to show the major components, how data flows between them, and what technologies were used, giving anyone a clear overview of the entire structure.

### Code Documentation

This is the explanation embedded within or alongside the software code itself. It includes comments that clarify complex logic, instructions for setting up the project, and detailed descriptions of how other programmers should interact with the system's functions.

### Model Documentation

This is specific to the machine learning model. A "Model Fact Sheet" is a common format that summarizes the model's purpose, the data it was trained on, its performance strengths and weaknesses, and any known limitations, enabling informed and responsible use.

### Operational Documentation

This consists of the manuals for running the system. It includes step-by-step deployment instructions, explanations of configurable settings, and troubleshooting guides to help fix problems, ensuring the system can be maintained reliably.

## Process Documentation

This documentation captures the "why" behind the system—the decisions and methods used during its development. It includes records of which modeling approaches were tested, why a particular algorithm was chosen, and how various data processing decisions were made. This creates an invaluable historical record that prevents teams from repeating past mistakes and provides crucial context for future iterations. Ultimately, this practice turns development from a one-off project into a repeatable and auditable process, ensuring institutional knowledge is retained.

### Requirements Documentation

This defines what the system is supposed to do from the user's perspective. It outlines the desired features and the criteria that must be met for the project to be considered successful, guiding the entire design and testing process.

### Design Documentation

This records the key decisions made during the system's creation. It explains why certain approaches were chosen over others and what trade-offs were considered, preserving the reasoning that is often forgotten over time.

### Experiment Logs

These are the detailed notes from the model development phase. They track every experiment—what was tried, what worked, and what failed—preventing teams from repeating past mistakes and providing context for the current model design.

### Change Logs

This is a history book for the system. It meticulously records what was changed in each update and the reason for the change, which is invaluable for understanding the system's evolution and diagnosing new issues. This creates a clear audit trail that simplifies troubleshooting when a new update introduces unexpected behavior.

## User Documentation

This is created to help end-users understand and effectively use the system. It typically includes clear instructions, practical examples of inputs and outputs, and answers to frequently asked questions. Well-designed user documentation reduces support requests and empowers users to leverage the system's capabilities to their full potential, thereby increasing overall adoption and satisfaction.

### User Guides

These are instructional materials that explain the system's features. They often include a "Getting Started" guide, detailed feature descriptions, and step-by-step tutorials to help users achieve their goals. Their primary purpose is to empower users, enabling them to become self-sufficient and productive with the tool.

### Best Practices

This documentation offers expert advice on how to get the best results from the system. It guides users on when and how to use certain features, what the limitations are, and what common pitfalls to avoid. Following these recommendations helps ensure users achieve consistent, high-quality outcomes and avoid unintended consequences.

### FAQ Documents

These address the most common questions and concerns users have. They are a dynamic resource that evolves based on real user feedback, helping to reduce the burden on support teams. A well-maintained FAQ provides immediate answers, improving the user experience and allowing experts to focus on more complex issues.

### Release Notes

These announcements inform users about updates. They clearly communicate new features, bug fixes, and any known issues in the latest version, helping users adapt to an evolving system. Transparent release notes build trust by keeping users informed and prepared for changes that might affect their workflow.

# Compliance and Audit Documentation

This documentation provides evidence that the system adheres to regulations and ethical standards. It typically includes detailed records of data lineage, model validation results, fairness audits, and user consent management. Maintaining this documentation is not optional; it is essential for passing formal regulatory reviews and demonstrating due diligence. A well-organized audit trail can significantly streamline the compliance process and protect the organization in the event of a legal challenge.

## Privacy Documentation

This records how user data is handled. It specifies what data is collected, how it is used, who has access to it, and when it is deleted, demonstrating compliance with privacy laws. This documentation serves as a critical reference during internal reviews and external audits by regulators.

## Security Documentation

This describes the measures in place to protect the system and its data from threats. It outlines the security controls and protocols, showing that the organization has exercised due diligence. This is often a foundational requirement for achieving industry certifications and assuring partners of the system's integrity.

## Fairness Documentation

This serves as documented evidence of the steps taken to ensure system fairness. It includes records of bias testing, mitigation strategies, and their outcomes, reinforcing accountability and transparency. Maintaining such records builds trust among stakeholders and the public alike.

## Audit Trails

These are automated logs that record system activity. They provide a definitive record of who did what and when, which is essential for investigating incidents and ensuring accountability. Tamper-proof audit trails, records that show who did what and when, are a key part of strong security and compliance.

# 6.11 Conclusion

Building robust machine learning systems requires far more than training accurate models. It demands a systematic, interdisciplinary process that spans the entire lifecycle–from problem definition through ongoing maintenance and evolution. This holistic approach forms the foundation for creating systems that are not only technically advanced but also genuinely valuable, trustworthy, and sustainable.

## Key Principles for Success

The journey to a successful machine learning system is guided by several enduring principles:

- **A Comprehensive, Iterative Process:** Success depends on every stage–from data collection and model training to deployment, monitoring, and feedback integration. The lifecycle is inherently cyclical: initial deployments serve as a foundation for continuous measurement, learning, and improvement.
- **Beyond Technical Excellence:** Production-ready systems must balance competing demands. They must scale efficiently, safeguard privacy and security, uphold fairness and transparency, and deliver intuitive user experiences that build trust.
- **Foundational Documentation:** Comprehensive documentation is not an afterthought. It enables understanding, accountability, collaboration, and long-term system reliability.

## The Interdisciplinary Nature of Machine Learning

The development of effective machine learning systems is inherently collaborative., requiring integrating several critical perspectives:

- **Technical and Domain Expertise:** Algorithmic insight must merge with deep domain understanding to solve meaningful problems effectively.
- **Human-Centered Design:** User research and thoughtful interface design must ensure systems are understandable, trustworthy, and truly usable.
- **Ethical and Business Alignment:** Ethical reasoning must guide design choices, ensuring fairness and accountability while linking technical capability to organizational value and social impact.

## From Potential to Value

As machine learning shapes both critical decisions and everyday life, the quality of the development process becomes paramount:

- **The Reward of Rigor:** Well-engineered systems built through disciplined, transparent processes deliver meaningful benefits while managing risk responsibly.
- **The Cost of Neglect:** Poorly developed systems can amplify bias, erode privacy, and cause unintended harm—often due to narrow focus on model metrics alone.

The framework presented in this chapter offers a foundation for navigating this complexity. Though specific tools and techniques will continue to evolve, the enduring commitment to systematic development, rigorous evaluation, responsible deployment, and continuous improvement remains the surest path from algorithmic potential to real-world value.

As machine learning matures, the challenge shifts from creating powerful algorithms to ensuring that systems are reliable, understandable, and aligned with human values. The next frontier lies not in accuracy alone, but in building ecosystems where technical performance, ethical integrity, and human trust coexist. Organizations that master this balance will define the future of responsible AI.

# Index

## E



## F



## G



## H



## I



## K



## L

## T



## U



## V

# About the Author

Arshad Khan is an analytics and artificial intelligence professional with over four decades of experience spanning consulting, corporate leadership, and education. He has worked with or advised leading organizations including Accenture, PwC, Deloitte, Chevron, Bose Corporation, Chrysler, Genentech, PepsiCo, Hitachi-America, and numerous other Fortune 500 companies.

Mr. Khan served as Program Manager for a major AI initiative in Silicon Valley, where he led efforts to design and implement advanced data and machine learning solutions. This hands-on leadership role deepened his expertise in AI strategy, deployment, and governance.

A prolific author, he has written twenty books, including seven on business intelligence and analytics and three on artificial intelligence. His publications are widely used by professionals and students for their clarity, practical focus, and depth of insight.

As an adjunct professor for more than 35 years, Mr. Khan has taught data visualization, analytics, and artificial intelligence at ten universities, including Golden Gate University, Santa Clara University, and the University of California extensions at Berkeley, Santa Cruz, and San Diego.

He holds a graduate degree in chemical engineering and an MBA, combining technical rigor with business acumen. Throughout his career, Mr. Khan has bridged the worlds of technology, education, and applied analytics—helping organizations and individuals harness the power of data and AI responsibly and effectively.

Made in the USA
Monee, IL
04 November 2026

67382350R00075